WITHIN REASON

WITHIN REASON

Rationality and Human Behavior

DONALD B. CALNE

PANTHEON BOOKS

NEW YORK

All rights reserved under International and Pan-American Copyright
Conventions. Published in the United States by Pantheon Books, a division of
Random House, Inc., New York, and simultaneously in Canada by Random
House of Canada Limited, Toronto.

Pantheon Books and colophon are registered trademarks of
Random House, Inc.

Grateful acknowledgment is made to The MIT Press for permission to reprint
the translation of a poem by Sappho from *The Rationality of Emotion* by
Ronald de Sousa. Copyright © 1987 by The Massachusetts Institute of
Technology. Reprinted by permission of The MIT Press.

Library of Congress Cataloging-in-Publication Data

Calne, Donald B. (Donald Brian)
Within reason : rationality and human behavior / Donald B. Calne.
p.   cm.
Includes bibliographical references and index.
ISBN 0-375-40351-5
1. Reason. 2. Reasoning (Psychology) I. Title.
BF442.C35 1999
153.4'3—dc21   98-38230
CIP

Random House Web Address: www.randomhouse.com

Book design by Julie Duquet

Printed in the United States of America
First Edition
2 4 6 8 9 7 5 3 1

To my wife, Susan

CONTENTS

ACKNOWLEDGMENTS

First, I thank Oliver Sacks, the writer who brought the excitement of neurology to the general public; he kindly looked at the manuscript and gave me encouragement. Beverly Goldberg guided me through the maze of possible publishers. Michael Schulzer, Dan Overmeyer, Edward Hundert, and Tony Phillips spent much time in reviewing the text. Many colleagues commented on the chapters in their special areas of expertise, in particular Braxton Alfred, David Crockett, Serge Guilbaut, Reinhard Horowski, Trevor Hurwitz, Ashok Kotwal, Peter Liddle, Victoria McGeer, Bill McKellin, Hugh McLennan, Jeff Miller, Chris Pallis, Jon Stoessl, Michael Tenzer, François Vingerhoets, and Earl Winkler. Elsie Wollaston gave bibliographic assistance. Iain MacPhail, John Nichol, Paul Nicholson, Janet Pau, Michael Pitfield, and Gordon Vichert provided constructive advice, and my son, Max Calne, shared many useful thoughts. I would also like to thank Peter Dimock of Pantheon Books, whose wisdom and editorial experience were combined with unusual warmth, so that our relationship became one of friendship and mutual respect. Last and most importantly, I thank my wife, Susan, for her consistent help; above all, I thank her for enduring my distraction with the book over many evenings, weekends, and vacations.

PREFACE

This book looks at reason and how it affects our lives. Where does reason come from, what is it, what can it do, and what can it not do? How do our brains provide us with minds with which we create concepts of space, time, and causation, and how have these, combined with the capacity of memory, enabled us to establish the faculty of reason? I will argue that reason has grown beyond its innately humble origins to acquire a seemingly invincible power that encroaches upon and overshadows the rest of life. Paradoxically, one of reason's greatest contributions may yet turn out to be precise delineation of its own limits, thereby making possible greater responsibility on the part of the planet's dominant species. The task of this book is to remind us that reason is a biological product, with biological purposes and biological limitations. We are motivated by instinctive urges and emotions linked to cultural forces—reason is their servant and not their master.

Would the world be a better place if reason were somehow put in the driver's seat? Will it become possible to insert reason into the mechanisms of human motivation, the way molecular biologists can insert a gene into the decision-making processes of a cell? Is language essential for the exercise of reason? What role does reason play in morality? These are some of the questions we must ask in order to approach the larger issue of the part played by reason in human life.

WITHIN REASON

1

INTRODUCTION

*The sordid and savage story of history has been written by
man's irrationality, and the thin precarious crust of
civilization which has from time to time been built over the
bloody mess has always been built by reason.*
—BARBARA WOOTTON*

WHEN I WAS young, I was taught that education was important because without it we would be doomed to stupid behavior and opinions based upon prejudice. Educated people, I was told, were able to make wise decisions and to distinguish between right and wrong because they had the power of reason available to them. I grew up during the Second World War, and one of my most vivid childhood memories was a kaleidoscope of movie images shown in British schools in 1946, depicting the Nazi death camps. I was puzzled, for I had been led to believe that these atrocities were the irrational work of barbarians whose evil stemmed from ignorance. Yet I was also taught that Germany was distinguished by a history of outstanding reason, expressed in philosophy, music, poetry, science, and technology. If what I had been told about wisdom coming from ed-

---

* *Testament for Social Science* (London: Allen and Unwin, 1950). Barbara Wootton, the daughter of a Cambridge don, worked for the British Labour Party between the wars. Later, she became a magistrate and a baroness.

3

ucation was true, how was it possible that Germany, the home of
Bach, Beethoven, Brahms, Goethe, Leibniz, and Kant, could become
a nation driven by hatred and complicit in the worst crimes against
humanity that the world had ever seen? The conflict remained with
me and gradually matured into a series of questions. Does reason direct what we do? If we think more, do we behave better? In short,
could the nightmare of the Second World War have been avoided if
the leaders of National Socialism had acquired, in some miraculous
way, a sudden capacity for more reason?

Sadly, the facts do not support this. The intellectuals of Germany
were among the first to embrace National Socialism. Wagner and
Nietzsche blazed the trail in the nineteenth century, and by 1933 large
numbers of university faculty were ready to champion National Socialist ideology. Other representatives of the educated classes, the
lawyers and the physicians, and even more practical leaders, the in-
dustrialists, joined the throng. Many Europeans outside Germany
looked on with approval. In some respects Hitler was expressing a
widespread and influential sentiment that permeated the thinking of
European intellectuals.* The National Socialist movement was not
conceived by ignorant people; its roots lay in the intelligentsia. It is
difficult to escape the conclusion that if the leaders of National Socialism had achieved a dramatic increase in their capacity for reason-

---

* In *The Intellectuals and the Masses* (London: Faber and Faber, 1992), John Carey provides a persuasive analysis of the elitist attitudes among the English literary intelligentsia—even when they publicly professed liberal views. In Germany there was a long
and highly influential tradition of ultranationalist ideology in student fraternities and in
the established intellectual community. Forty-five percent of German physicians were
members of the Nazi party, at a time when membership among the total population was
fifteen percent. See M. H. Kater, *Doctors under Hitler* (Chapel Hill: University Press, 1989).
Konrad Lorenz joined the Nazi party after Austria was incorporated into Germany in
1938, and he was active on behalf of the "Racial Political Office." Karl von Frisch helped
the Goebbels Military Fund. Lorenz and von Frisch, both distinguished biologists,
received the Nobel Prize together in 1973. See P. Weindling, *Health, Race and German
Politics between National Unification and Nazism, 1870–1945* (Cambridge: Cambridge
University Press, 1989). See also U. Deichmann, *Biologists under Hitler,* translated by
Thomas Dunlap (Cambridge, Mass.: Harvard University Press, 1966).

ing in 1939, their regime would simply have pursued its policies with a more intelligent war machine; the goals would not have changed. National aggrandizement, territorial expansion, and institutionalized racism would have continued with more efficient weapons. The management of the "final solution of the Jewish question" was entirely dependent upon the ability to harness a product of reason—modern technology—to the problem of mass transportation, the safe manufacture and containment of Zyklon B, and the engineering of incinerators that could be fueled by the continuous ignition of melting human tissues.

Yet there was a paradox: vital resources were committed to the "final solution" though this diversion of manpower and material weakened the war effort. The German armies were in full retreat on all fronts. How can one make sense of this subversion of effort? The assembly of historical facts seems to amount to a grotesque concoction of reason with unreason. These events of my childhood kindled a deep personal curiosity about why people do things. The curiosity stayed with me, and as the war receded, my questions took a more general form. Have people all over the world been so irrational throughout history, and if so, why? I was assuming a natural tendency for us to be rational, but where is the evidence for this? I rephrased the questions into a more approachable form: What do we know about reason and the way we use it?

To pursue my curiosity, I began to look for information. Where possible, I have taken the position of a neurologist, because I happen to be a neurologist, but also because neurology gives a close-up view of some of nature's harshest but most illuminating experiments—injuries that damage different parts of the brain. Strokes, tumors, and trauma may isolate and destroy one particular mental capacity, leaving everything else intact. Nature's disasters can thus help to explain how the brain works and shed some light on the instrument that stands at the center of the intellect: human reason.

I have taken to heart Francis Bacon's dictum that "truth emerges more readily from error than from confusion" *(1)*.* Where evidence is indefinite, I have tried to find the most probable position, recognizing that it may ultimately prove to be wrong. My exploration has led me to a conclusion that was somewhat unexpected. Reason is a biological product—a tool whose power is inherently and substantially restricted. It has improved *how* we do things; it has not changed *why* we do things. Reason has generated knowledge enabling us to fly around the world in less than two days. Yet we still travel for the same purposes that drove our ancient ancestors—commerce, conquest, religion, romance, curiosity, or escape from overcrowding, poverty, and persecution.

To deny that reason has a role in setting our goals seems, at first, rather odd. A personal decision to go on a diet or take more exercise appears to be based upon reason. The same might be said for a government decision to raise taxes or sign a trade treaty. But reason is only contributing to the "how" portion of these decisions; the more fundamental "why" element, for all of these examples, is driven by instinctive self-preservation, emotional needs, and cultural attitudes. We are usually reluctant to admit the extent to which these forces govern our behavior, and accordingly we often recruit reason to explain and justify our actions. The transparency of our efforts is revealed by the term we have coined for covering up this irrational behavior: rationalization.

The rest of this chapter outlines some of the hopes for reason and some of the disappointments. In chapter 2 I develop a working definition of "reason." I deliberately avoid concentrating on any one aspect of reasoning, such as the psychological process of making decisions. If I pursued a narrow area in depth, I would thwart my attempt to view the whole picture of reason's place in human life, so in subsequent chapters I have included a wide range of topics, taking

---

* Most of the important works referred to in this book are designated in the text by an italic numeral in parentheses. Full information on the work is given at that numeral in the chapter-by-chapter Works Cited section at the back of the book. Alternatively, some sources are provided with a comment, in a footnote.

each "to only one decimal place" to keep the totality of rational behavior in perspective.

Since reason and language are so closely linked, I devote chapter 3 to delving into linguistics and the evidence that under exceptional circumstances neurological disease can separate reason from language. In chapter 4 I trace the development of social organization, for there is now a body of evidence suggesting that intelligence evolved from the need for individuals to cooperate in order to survive. In chapters 5 through 11, I follow the role of reason in the creation and maintenance of our most hallowed institutions: morality, commerce, government, religion, art, and science. In the remaining chapters I look at how reason operates in the human brain and mind.

Reason, like instinct and emotion, has evolved to facilitate the attainment of biological goals. Curiously, we have often found it easy to use reason in a harmful way. Chekhov's prophetic words, written a century ago, have a contemporary ring: "Man has been endowed with reason, with the power to create, so that he can add to what he's been given. But up to now he hasn't been a creator, only a destroyer. Forests keep disappearing, rivers dry up, wild life's become extinct, the climate's ruined, and the land grows poorer and uglier every day."*

We proclaim that the disastrous events of the two world wars will not be repeated, but the same forces that led to those tragedies persist today, barely beneath the surface. Contemporary examples are, unfortunately, abundant: the death of 30 million Chinese as a result of political mismanagement committed under the ironic slogan of "The Great Leap Forward," the killing fields in Cambodia, the slaughter of Kurds in Iraq, the "ethnic cleansing" in Bosnia, and the tribal massacres in Rwanda. The list is long and continues to grow.

---

* These words are spoken by Astrov, a physician (like Chekhov) and a keen environmentalist in *Uncle Vanya*. Astrov goes on to drive home his point: "I've saved all these woods. When I listen to the wind in those young trees that I've planted myself—it seems to me that I do have some power over the climate. If a thousand years from now the world is a happier place, who knows, perhaps I'll have had something to do with it." *Uncle Vanya* was first performed in 1899, five years before Chekhov's death.

The evidence compels the conclusion that in spite of our capacity for reason, we remain tied to the motivation provided by our biological drives and cultural attitudes. In these circumstances I argue a humanist position informed, even guided, by recognition of the limits of reason. To place reason in perspective, we should take it down from the pedestal upon which expectations of supremacy have placed it. When we do this, we find that in many ways reason is like language, for both are highly complex instruments developed for biological purposes. They help us to achieve what we want, without having any real impact on why we decide what we want. Both operate unobtrusively; we take both for granted.

HOPE AND DISAPPOINTMENT

In the past, reason has been given the status of an independent, external and ultimate authority, with the ability to confer wisdom and goodness. Like a deity, reason was conceived as all-powerful. The ascent of reason began when the ancient Greeks surveyed the universe and attempted to sort out the confusion of ideas that had accumulated over previously known history. The Greeks were not the first to pay attention to reason, but they used it more extensively than anyone had before, raising rational discourse to an exalted status. Sophocles caught the spirit of his times in a single line: "Reason is god's crowning gift to man" (2). Aristotle echoed this view a century later: "For man, therefore, the life according to reason is best and most pleasant, since reason more than anything else is man" (3). In Rome, Cicero proclaimed that "reason is the ruler and queen of all things" (4). Similar views were forming in India, and in China Confucius was on the same track.*

---

* The eruption of new ideas during this era, 500–1 B.C., can be seen as a first Age of Reason. There was an explosion of philosophy, literature, and mathematics; major religious changes culminated later in the appearance of Christianity and Islam—and these all had significant political consequences. A second Age of Reason took place in the East, led by Arab mathematicians, from the seventh to the fifteenth century. For the third Age of Reason, the action returned to Europe in the seventeenth century; it was like a replay of the first, leading to equally spectacular developments in philosophy, art, science, and politics.

This early optimism proved to be transient. After a few centuries of achievement, the first Age of Reason went into a prolonged decline in Europe. Knowledge became a product generated from *a priori* principles; it was divorced from observation, but it carried the authority of unquestionable certainty. Intellectual innovation in the West slowed down for over fifteen hundred years, although in the East reason was burgeoning. The Islamic world made notable advances in mathematics, astronomy, medicine, and architecture. Akbar the Great, Moghul emperor of India from 1560 to 1605, declared, "The superiority of man rests on the jewel of reason" *(5)*.

Commerce and the Reformation weakened the traditional power of the church and the monarchies in Europe, and the great minds of the Renaissance broke through the mental barriers imposed over the Middle Ages. In England, Shakespeare planted reason firmly in the thoughts of Hamlet:

> Sure, he that made us with such large discourse,
> Looking before and after, gave us not
> That capability and god-like reason,
> To rust in us unused.
> (Act IV, sc. 4) *(6)*

Once released in Europe, reason leapt forward. It drove science, art, and literature; during the seventeenth and eighteenth centuries the surge of intellectual innovation and critical inquiry gave a distinctive name to the epoch, the Enlightenment. The tone was set by the "stern pursuit of accurate knowledge based on evidence, logic, and probability in preference to the colourful confusion of myth and legend that had satisfied a less critical age" *(7)*. In science, reason *demanded* linkage between observation and theory, and this gave a new order to the world. In the words of Isaac Newton: "Science consists in discovering the frame and operations of Nature, and reducing them, as far as may be, to general rules and laws—establishing these rules by observations and experiments, and thence deducing

the causes and effects of things."* Experiments were designed to test
hypotheses; if the hypotheses could not be disproved, they were incorporated into the growing body of knowledge and applied—to
engineering, architecture, medicine, and back into science. Émile
Borel depicted the excitement of a people testing a tool of immense
power: "The real inspiration of this splendid epic, the conquest of
the world by man, is the faith in human reason, the conviction that
the world is not ruled by blind gods or laws of chance, but by rational laws."†

Startling advances in understanding how the physical world
worked were accompanied by social upheavals. In political philosophy, Baruch Spinoza asserted that the purpose of the state was "to
lead men to live by, and to exercise, a free reason; that they may not
waste their strength in hatred, anger and guile, nor act unfairly
toward one another" (8). New political ideas erupted to produce
reforms that rocked the foundations of traditional dynasties.
Monarchies were replaced by democracies and the Industrial Revolution fueled the turmoil. The face of European society was transformed. As a result of these changes, faith in reason reached its zenith
toward the end of the nineteenth century. For the Victorians, anything was possible. Time and again reason "worked," so for many it
became a new god, possessed of great powers and intrinsic virtue.
The most direct expression of reason was science, which seemed invincible. Science and reason together would rid the world of poverty,
disease, and ignorance; they would vanquish prejudice and superstition; they would lead to a coherent explanation for everything
under—and beyond—the sun.

The hopes were not fulfilled. In the twentieth century, two devas-

---

* *Philosophiae Naturalis Principia Mathematica* (1687), otherwise known as the *Principia*, is one of the most important books in the history of science. In the words of Alexander Pope: Nature and Nature's laws lay hid in night.
                   God said, *Let Newton be!* and all was light.
† *Le Hasard* (Paris: Alcan, 1938). Borel, a professor at the Sorbonne, was a mathematician,
politician and journalist.

tating world wars, numerous small wars, and recurrent economic instability sapped confidence and optimism. The pendulum began to swing against reason and now the opposition is coalescing. The recoil from reason takes on the aspect of a surreal motion picture. The growing strength of cults, religious fundamentalism, and political extremism reflects this disenchantment. Unreason flourishes with the rise and increasing popular authority of clairvoyants, spiritualists, astrologers, faith healers, devotees of alternative medicine, and new age extraterrestrial communicators. These exponents of unreason are irrational because they reject, deny, or misinterpret relevant information that is available through observation. Widespread anti-intellectual forces denounce science as a regressive influence driving imperialism and militarism—even sexism and racism. A new and fashionable view holds that science is a subjective, culturally determined ideology with nothing "real" behind it. The letter of invitation to the Nobel Conference XXV, held in 1989, warns: "As we study our world today, there is an uneasy feeling that we have come to the end of science, that science, as a unified, universal, objective endeavor, is over."* The problem is not confined to science; there is a fragmentation of public support for all academic activity. Governments have lost interest in the university and its potential. The Chinese Cultural Revolution showed how easily political forces can exploit anti-intellectual sentiments into a massive popular movement capable of destroying art, science, and medicine. The onslaught against reason in China was all the more alarming because it achieved such sweeping success in a nation whose historical roots are steeped in art, science, and medicine—whose people were pioneers of reason.

Why have so many turned against reason? There are several explanations, but among the foremost must be failure of the quixotic hopes vested in it. Reason was misrepresented as an all-powerful, divine force, with its own supreme mission. In fact, it has

---

* Cited in Gerald Holton's *Science and Anti-Science* (Cambridge, Mass.: Harvard University Press, 1993).

no aim and no inherent goodness. Reason is simply and solely a tool, without any legitimate claim to moral content. It is a biological product fashioned for us by the process of evolution, to help us survive in an inhospitable and unpredictable physical environment. It is also a tool to enable us to compete with other animals that are larger, faster, and stronger, with longer claws and more powerful jaws.

These may seem hostile allegations—that reason is, by its nature, always constrained to provide a service rather than set a policy—but I do not intend to slight or defame reason. Indeed, my strongest argument is that we must make every effort to exercise and preserve the faculty of reason. This task has become a challenge because reason has been discredited by exaggerated claims and false hopes. Those who oppose reason have no difficulty in pointing to its failures, but these failures have all stemmed from misguided expectations. The hopes sprang from natural, if reckless, optimism—without critical thought about the nature of reason and without recognizing the need to clarify what reason can and cannot do.

The record needs to be set straight. To this end we can start by asking what reason is, where it came from, and what it does. As we begin to examine reason, we soon feel the breadth of its influence in our lives. A microscope, concentrating on one small area, does not have an adequate field of vision. A wide-angle lens must be used to gain perspective. We need to survey an extensive array of human endeavors in order to discern the full extent of reason's presence in human life. The evidence I cite will inescapably be far-ranging but not, I hope, flawed on that account.

As the book's chapter titles indicate, my survey is indeed broad—some might argue too broad to say anything new or significant. But my purpose will have been served if, in the course of this book, the reader comes to accept its central, all-important, though perhaps deeply counterintuitive argument: that the mental faculty of reason is a real, specifiable, and indispensable human capability active in most domains of human existence *but* that it cannot assign or control the purposes to which it is put. I look at a wide assortment of human endeavors to show both rationality's functioning within each domain and simultaneously to note its inability to supply the purpose of the power it confers. My conclusion is easily stated but difficult to put into practice. I argue both that human reason must be vigorously defended against the growing forces of unreason in the contemporary world and, simultaneously, that the reliance upon it to supply the content of human motivation must be given up. Only such double awareness of the human faculty of reason can prevent the disenchantments and resentments that inevitably result from misplaced hope.

2

A DEFINITION OF REASON

Reason is wholly instrumental. It cannot tell us where to go;
at best it can tell us how to get there. It is a gun for hire
that can be employed in the service of any goals we have,
good or bad.

—HERBERT A. SIMON*

W HAT, EXACTLY, IS reason? To a teacher, it is an intellectual exercise for developing the minds of the young. To a lawyer, it is a way to confirm or refute testimony. To an economist, it is a means of allocating resources to maximize efficiency, utility, and wealth. To a scientist, it is a method for designing experiments to explore hypotheses. A dictionary, which provides a consensus of how most people use words, offers a broad definition: "The mental faculty (usually regarded as characteristic of humankind, but sometimes also attributed in a certain degree to animals) which is used in adapting thought or action to some end" (1). Nicholas Rescher adds: "To behave rationally is to make use of one's intelligence to figure out the

---

* *Reason in Human Affairs* (Stanford, Calif.: Stanford University Press, 1983). In this book Simon sketches out and refutes an "unwarranted optimism" concerning the use of reason. In 1978 Simon was awarded the Nobel Prize for "pioneering research into the decision-making process in economic organization."

best thing to do in the circumstances." Reason affords "the optimal chances of success in realizing one's appropriate ends" *(2)*. These concepts offer a roughly hewn definition, but one that provides a firm foundation on which to build an argument.

The principles by which reason operates are the domain of philosophy and psychology. But if we concentrate all our effort in such specialized areas, we shall become bogged down in esoteric technical discussions—over issues such as the rules of inference and theories of consciousness—and we will miss the big picture. Another human skill, manual dexterity, offers a useful analogy. Human hands are capable of astonishing precision in coordinating fine finger movements. Yet a deep understanding of manual dexterity does not come from studying the mechanics of how muscles move finger joints. More insight is gained by considering how the forelimbs were used for weight bearing in our distant ancestors, and how hands developed for grasping the branches of trees. The coarse manual movements needed for climbing became delicate and more effective in gathering food and fashioning tools. Skillful hand movements enabled our culture to produce watchmakers, violinists, and neurosurgeons.

We cannot be sure of the evolutionary origin of reason, but there is much evidence to suggest that it developed out of the capacity to react appropriately to complex, changing social situations—in a setting where survival of the species depended upon living in social groups. From studies on patients with disorders involving the front of the brain, we know that damage to the frontal lobe leads to loss of social skills and, in addition, that portions of the frontal lobes play a key role in what we might loosely call the process of making rational decisions. In neurological terminology, a major component of reasoning takes place in a frontal module of the brain, a module being a specialized, relatively independent group of interconnected nerve cells performing a particular category of tasks. Patients who have an injury confined to this module, caused by stroke, tumor, or trauma, are not confused or demented. They have no hallucinations or delusions. They have a more specific syndrome in which reason is disturbed without a generalized disorder of cognition. Appropriately enough, this isolated impairment of the module for reason is called "the dysexecutive syndrome," a condition analogous to selective disorders of language (the aphasias) or memory (the amnesias).

Patients with the dysexecutive syndrome have difficulty steering themselves through the ambiguities of daily life; they are no longer "street smart," for they cannot adjust to the rapidly moving patterns of physical events and human interaction around them, and they make mistakes in assembling rational sequences of thought. For example, one patient with the dysexecutive syndrome concluded that there are money-trees because "money is green [in the United States] and so are trees, so money must grow on trees." She was equally unable to put together a rationally organized series of actions—she would do her laundry by placing the box of detergent in the washing machine with her clothes.*

Other functions of the brain are necessary for the module of reason to operate. There must be consciousness, memory, and an ability to create and transform mental symbols. But these capabilities are not enough by themselves to find the best ways to reach goals. Our patient with the dysexecutive syndrome knew that money is green and that trees are green; she knew that clothes have to be washed and that detergent is used in washing machines. She just could not combine her concepts effectively to solve problems: where to get money; how to wash clothes. Why is her disorder termed the dysexecutive syndrome? An executive, according to *Webster's New Collegiate Dictionary,* is "one who holds a position of administrative or managerial responsibility," and this is the role of

---

* This example of a patient with the dysexecutive syndrome is taken from J. D. Duffy and J. J. Campbell, "The Regional Prefrontal Syndromes: A Theoretical and Clinical Overview." *Journal of Neuropsychiatry and Clinical Neurosciences* 6 (1994): 379–387.

reason when it comes to solving problems. In a corporation, the board of directors set the goals and the chief executive officer finds the best way to reach them. If we can, individually, be looked upon as micro-corporations, we each have a board of directors—the ensemble of instincts, emotions, and cultural attitudes that make up motivation—and we each have a chief executive officer—the faculty of reason.

One thorny problem we have to acknowledge in discussing definitions is the difficulty of dealing with the term "intelligence." Most people regard intelligence as intellectual capacity—it is hard to reach a definition with greater precision.* If intelligence is the full breadth of intellectual capacity, reason is one of its constituent parts. An analogy helps to illustrate the point. Voluntary movement is an important, broadly defined neurological function, so we can return to consider manual dexterity. When a patient has a neurological problem with using a hand, diagnosis requires a breakdown of movement into its various components. First, the brain has to receive information on where the fingers are; this is tested by asking the patient to identify small displacements when the examiner moves the subject's finger joints (visual clues are excluded by closing the eyes). Then there must be muscle strength; this is tested by asking the patient to press the fingers against resistance. There must also be coordination; this is tested by asking the patient to perform rapidly repetitive movements. Finally, there is the task of assembling a series of movements to achieve a goal; this is tested by asking the patient to put a piece of paper in an envelope. Diseases involving different parts of

---

* Measures of intelligence were designed for schoolchildren by Alfred Binet, who became director of physiological psychology at the Sorbonne in 1892. The famous battery of tests, which became known as the Binet Scale, were requested by the French government for the selection of children for schools. Binet introduced the concept of the Intelligence Quotient—the mental age divided by the chronological age, multiplied by 100. Subsequently, many problems arose with IQ testing because the results are distorted so much by cultural variables. Now measurements of IQ are interpreted with great caution.

the brain can selectively take out any one of these elements of movement. In this analogy, the overall capacity for movement is equivalent to intelligence; the preliminary requirements of sensation,
muscle power, and coordination correspond to consciousness, memory, and the ability to transform symbols. The ability to design an
optimal series of movements to reach a goal is equivalent to the faculty of reason.*

Reason gives a flexible, infinitely adjustable range of reactions to
changing circumstances. It is a deliberating process of working
things out to solve problems; once the problems have been worked
out, the solutions can be stored in our memories so we can call up
the same solution to the same problem whenever necessary, without
repeatedly resorting to reason. Thus reason is a means for improving
our ability to survive in the face of a challenging environment. Since
we are now dealing with definitions, we should specify how we will
use the term "survival." Are we referring to the survival of genes, individuals, the species, or the culture? In this book, unless otherwise
stated, survival will denote persisting or increasing "person-years"—
"person-years" being the product obtained by multiplying the number of people with their life span. This definition takes into account
the decision of some communities to use birth control to limit overcrowding and extend life expectancy: thus a small number of people
living a long time becomes equivalent to a large number of people
living a short time.†

---

* Failure of the ability to assemble a series of movements is called apraxia, and apraxia
can be just as disabling as loss of muscle power. The dysexecutive syndrome is, in a sense,
an apraxia of concepts.

† Different definitions of survival have different biological implications. For example, if
we were interested in reproductive capacity and sustaining the species, a relative increase
in the young adult population would lead to more children and therefore more genetic
variation. For us, however, the most important effects of genetic variation were expressed
before *Homo sapiens* emerged. The definition we have chosen gives a compromise index
of human success, for just counting heads would fail to acknowledge the impact of the increase in average duration of life achieved by reducing infective disease and famine.

Reason is a tool that has evolved through natural selection over millions of years; it has been constructed by living organisms, for living organisms. The biological benefits of reason have contributed to the natural selection of our distant hominid ancestors, but as far as we know brain power has not changed since *Homo sapiens* first appeared. Of course we can achieve much more now compared to *Homo sapiens* 200,000 years ago, but this ability has been gained by the accumulation of experience and knowledge—it has not been gained by developing sharper minds. Over the course of human history people have compiled and consolidated information about the laws that govern the way the world works, and it is the application of this accumulated wisdom that has, within the last few centuries, enabled us to dominate the planet.

Reason is a facilitator rather than an initiator; we engage reason to get what we want, not to choose what we want. Instinct drives us to seek food when we are hungry, and when food is scarce reason is used to devise a way of finding or producing it. So reason is a tool just as language and mathematics are tools. Reason, language, and mathematics share this humble, practical origin, yet through the buildup of our cultural "database," from generation to generation, they have been able to reach lofty abstract heights. Reason has been elevated to symbolic logic, language to metaphysical poetry, and mathematics to probability theory. Cultures can take simple biological processes and embellish them to such levels of sophistication that their original purpose becomes remote and easily forgotten.

REASON HAS ANOTHER important aspect—it generally "works." It is successful in helping us get what we want, and this is a key feature that distinguishes reason from unreason. For example, many of us want to know how long we will live. Our life insurance companies want to know this just as much. To obtain an answer, there are many options that ignore reason—astrology, palmistry, or gazing into crystal balls. These methods do not work, or more precisely, they work much less consistently than reason. For more effective prediction, reason must be brought to bear because, as we shall see, it gives a special priority to observation. A rational examination of the evidence tells us that life expectancy is influenced by a number of specific variables such as age, weight, gender, family medical history, occupation, diet, tobacco consumption, and alcohol intake. Blood tests and X rays may increase the accuracy of the estimate. Life insurance companies make large amounts of money by using reason, in the form of medical and statistical science, to construct actuarial tables. They assign priority to observation over theory. They do not hire fortune-tellers.

What distinguishes medical scientists from fortune-tellers? They both claim to have specialized knowledge, but they employ different methods. Fortune-tellers and faith healers operate irrationally because they are, in general, inconsistent and incoherent. In contrast, medical scientists use a rigorous system that has some distinctive features. By avoiding contradictions they become, in general, consistent. By staying in accord with other fields of knowledge they are coherent. They work things out by structured arguments that obey the rules of logic. In short, they engage the methods of reason.

The development of reason is likely to have been similar to the development of language. Both must have begun with simple elements that gradually took on more complex roles—reason assembling arguments, and grammar assembling sentences. Just as linguistic scientists have found universal features in language, philosophers have found universal features in reason. By a process termed "logical induction," *particular* observations lead to *general* conclusions, and the generalizations allow *predictions* to be made, so the future is inferred from the past. For example: When I threw a stick for my dog, Ochre, yesterday, he chased it (observation). Therefore, Ochre chases sticks (conclusion). So if I throw a stick for

Ochre tomorrow, he will chase it (prediction). Arguments are also constructed so that premises lead to conclusions by a process termed "logical deduction." For example: Dogs chase sticks (major premise). Ochre is a dog (minor premise). Therefore Ochre chases sticks (conclusion). Here the starting point may or may not be an observation, for the argument is equally valid if we observe dogs chasing sticks or if we define dogs as stick chasers.

While induction and deduction have a central role in reasoning, they are not the only players. Ockham's razor and Galileo's knife are instruments for cutting away the irrational. Ockham's dictum, also known as the principle of parsimony, advocates simplicity when testing theories: "Entities should not be multiplied unnecessarily."* In other words, if there are several possible hypotheses with similar merit, we should choose the simplest. Ockham's razor has often helped to advance knowledge—the choice of a complex hypothesis can be considered irrational when a simpler one is sufficient.

Galileo's knife gives precedence to observation over theory because observations can refute theories, and thereby tell us the real nature of the world *(4)*. Theories without observations can only tell us what *might be;* theories with observations tell us what *is.* Galileo's knife is a much more forceful instrument than Ockham's razor, yet it is so simple that it seems hardly worth mentioning. Galileo's knife is the principle that set free European intellects after the long span of the Middle Ages. Of course, not all theories can be tested by observation, and not all observations are accurate—any observation that is out of keeping with prevailing knowledge must be scrutinized very carefully before it is accepted. Although rabbits may be seen to come out of conjurors' hats, the conclusion that they live there is so inco-

---

* William of Ockham was an English Franciscan philosopher who lived from 1285 to 1347. He was excommunicated for heretical views and spent his later years under the protection of Louis of Bavaria, energetically promoting the separation of church and state. Ockham's razor is a dangerous instrument; while it is usually useful, it can sometimes lead to mistakes, so it has to be handled with care.

herent that it must be rejected. Galileo's knife has been indispensable
to experimental attempts to disprove theories—the essence of science.

The critical controversy, where Galileo applied the knife, was
the debate over whether the sun revolves around the earth, as tradition and the church dictated. Galileo gained fame and notoriety
by wielding his metaphorical knife fearlessly, at a critical time, on
a critical issue. Before and since Galileo, oppressive cultures have
imposed theories that are not in keeping with observations. Our
own "open society" might benefit from a few deft cuts with
Galileo's knife to dismember the prejudices underlying racism
and sexism—the assumptions of racism and sexism are cultural
theories that can readily be refuted by observation. For example,
the increasing success of women in highly competitive professions
such as medicine and the law shows how misguided traditional
views have been. But still now, as in Galileo's time, cultures impose attitudes about the world that contradict observations, and
people accept these attitudes even though they fly in the face of
reason.

IN ALL ITS applications, reason is efficient—rational thoughts and
actions are economical. For example, many routes are possible for
a traveler going from Vancouver to New York for a family gathering. The choice is straightforward when everyone is agreed on
how to measure cost and benefit. The purchase price of the ticket
is the major variable, so it is clear that the rational choice is the
cheapest.

But the situation may not be so simple, in which case the most
efficient choice may be elusive. The cheapest ticket may involve an
indirect flight with unnecessary waiting in airports. Different
individuals will weigh this inconvenience in different ways, so the
best decision will depend on what is most important for the particular traveler. The usefulness of the journey must be included in the

equation, and consideration will have to be given to the cost/benefit ratio for alternative journeys that might serve the same purpose. A whole system of mathematics, decision theory, has been assembled to deal with the task of choosing between different types of options. Normally, however, we do not resort to decision theory. Consciously or unconsciously, we assign values to the various relevant factors, and we make a choice.

When faced with complex or inadequate information, we fall back on a hybrid approach in which reason and emotion become intertwined. To illustrate this point, we can return to the task of choosing a flight from Vancouver to New York for a family reunion. We shall suppose an airliner has recently crashed, and this naturally leads to anxiety. Our decision must now take into account all sorts of predictions, based on the safety record of certain aircraft, certain airlines, and certain weather conditions. We now have two goals: a primary one, driven by a social instinct to visit family in New York, and a secondary one, driven by the emotional need to reduce anxiety.* So if we have cautious dispositions, we will probably choose a four-engine aircraft, owned by a prestigious airline, flying nonstop in good weather. If our emotional makeup is different we might choose the opposite, for our secondary goal could just as well be set by a craving for excitement.

We shall refer to a secondary emotional goal as a "gut reaction" because this term conveys the common experience of making a rather quick and easy decision that makes us feel good. In our usual activities, emotion, acting through a gut reaction, often prevails over reason. When we use gut reactions, we sometimes "cover up" by alluding to intuition, for gut reactions are generally less successful than reason. Yet gut reactions are not aberrations, for some

---

* Antonio Damasio has introduced the term "somatic markers" for these secondary goals. His hypothesis on the role of somatic markers is presented in *Descartes' Error: Emotion, Reason and the Human Brain* (New York: Grosset/Putnam, 1994).

decisions have to be made rapidly if they are to be effective. The ability to take quick action is useful when a delay would be risky. In our evolution, as in our present world, it is clearly helpful for us to have access to a repertoire of fast responses built from past biological, cultural and personal experiences. So emotion and reason each have a place in our lives, emotion for necessarily quick responses and reason when there is time for us to work things out.

George Santayana saw reason as having a role in "harmonizing" volatile instincts. This idea sits well with the evidence that reason evolved in close association with the development of social skills.

> Reason was born, as it has now been discovered, into a world already wonderfully organized, in which it found its precursor in what is called life, its seat in an animal body of unusual plasticity, and its function in rendering that body's volatile instincts and sensations harmonious with one another and with the outer world on which they depend. . . . Reason has thus supervened at the last stage of an adaptation which has long been carried on by irrational and even unconscious processes.*

These ideas can be folded into an explanation of how reason shapes ethics.

THE RATIONAL UNDERPINNING of morality involves repeated cost/benefit analyses. The needs of the individual and the community have to be reconciled—the best balance has to be found between competition and cooperation so that everyone can live together in a

---

* *The Life of Reason* (New York: C. Scribner's Sons, 1905–1906). Santayana's main interest was philosophy, but he was also a poet, novelist, and essayist. Born in Madrid, he spent much of his life outside Spain, mainly in the United States, England, and Italy.

society that has consistent and acceptable standards of behavior. The traditional means of achieving this reconciliation has been the development of ethical values, and, since the fundamental problems are the same in different communities, there is a core of universal ethical standards. The universality of basic moral positions is illustrated by the shared views of all the major religions and moral philosophies: they advocate the same controls on the same elements of selfish behavior. For a reasonable balance between the interests of the individual and the community, society must establish guidelines on physical violence, ownership of property, communications and sexual relationships—the corresponding commandments are not to kill, not to steal, and not to bear false witness, and not to commit adultery. These edicts set the ethical tone of most cultures, but values must have some flexibility to change with circumstances. Killing, for example, becomes an obligation rather than a vice in times of war. Similarly, new ethical issues arise as societies change—when we are confronted with ethical questions such as when to terminate the life support system of a terminal patient, the best answer is reached by the application of reason. We seek an optimal balance between the interests of the individual and the community, so, taking the newly determined action becomes "good," and failing to take it becomes "bad."

Morality is inherently concerned with resolving the conflict between competition and cooperation, but reason does not set this goal. The goal arises from the instinctive drive for *Homo sapiens* to be a social animal.

Yet some goals vary from individual to individual and from culture to culture. The voluntary decision to lay down one's life is an interesting response to a cultural imperative. The individual has to decide between the options of obeying or rejecting the demand for total self-sacrifice. The process of making the decision is usually driven by secondary goals, such as the need to avoid the emotion of shame. Sometimes the secondary goal is more positive—the quest for everlasting happiness, glory, or respect. For example, until the nineteenth century Fijian religion taught the wives of chiefs to seek their own strangulation or live burial when their husbands died. Their faith told them that (1) chiefs had a life after death; (2) in the afterlife, chiefs had a better time if their wives accompanied them; and (3) in the afterlife, the wives acquired higher status if they were with their husbands. None of these assertions was accessible to verification or falsification; they constituted religious dogma, but if the wife fully believed the dogma, she would choose to die with her husband.*

Religion is not the only source of cultural imperatives that can command self-destruction. In the Second World War, Soviet troops were mostly atheist but they gave up their lives as willingly as anyone else. The Battle of Berlin could have been fought carefully and slowly to conserve the lives of young Russians, but the campaign was a ferocious onslaught in which the Soviet Union bore heavy casualties because Stalin was in a political race with the Allies. The courage of the Red Army was driven by cultural forces that tapped deep feelings. A standard repertoire of emotions was harnessed, as for all wars, including hatred of the enemy and a lust for revenge. Cultures still pose the traditional highly charged options—honor or disgrace, the same alternatives that always face men and women in war.

REASON IS A powerful tool, but one that is unable to determine goals. How does it come about that reason is limited in this way? We must look at the nature of motivation to answer this question,

---

* The body of the wife—and sometimes there were several wives—would be oiled and clothed (including ornaments), and the face and breasts would be colored with vermilion and tumeric. She would then be laid down beside the great man. This description comes from Edward Burnett Tylor's classic *Primitive Culture* (London: J. Murray, 1871). Tylor had no university degree, but he became the first professor of anthropology at Oxford.

and then the key to the puzzle becomes disarmingly simple. Motivation is the drive to find mental rewards and to escape mental punishments. We want to do what will make us feel happy and satisfied, and we avoid what will make us feel sad and frustrated. Instinct and emotion motivate us because their satisfaction brings happiness, and their frustration brings dismay. Cultures can tie their goals to instincts and emotions—that is how they are able to motivate.

This concept of motivation can be illustrated by examples. We feel pleasure when we instinctively escape from danger, or when we emotionally reciprocate affection, or when we achieve a cultural goal such as winning an Oscar. In contrast, reason lacks the capacity to motivate because it cannot make us feel anything. Its nature does not include any direct link to mental rewards, although it is, of course, always available to be applied to a task that entails a reward. Logical statements, cost/benefit analyses, and scientific experiments are emotionally neutral. They only give psychological rewards when they are bound to goals such as prestige, which, in turn, satisfy emotions such as pride. In addition, the "facts" of reason can be manipulated by culture to generate the "feelings" of emotion. In George Orwell's *Animal Farm* the emotionally neutral facts that animals have four legs and humans have two become emotionally charged when the pigs orchestrate a campaign against Farmer Jones with the slogan "Four legs good, two legs bad."* In the same way culturally induced attitudes can give a mathematician pleasure from seeing reason engaged to solve an abstract theoretical problem. Yet this is a layer of emotion that has been

---

* *Animal Farm* was first published in 1945. "The humbler animals set to work to learn the new maxim by heart. FOUR LEGS GOOD, TWO LEGS BAD, was inscribed on the end wall of the barn. . . . When they had once got it by heart the sheep developed a great liking for this maxim, and often as they lay in the field they would all start bleating 'Four legs good, two legs bad! Four legs good, two legs bad!' and keep it up for hours on end, never growing tired of it."

artificially imposed. Reason remains remote from the warmth of human passion. Language reflects these essential differences. Passion has the dynamic energy to "fire," "burn," "stir," "move," "animate," "arouse," and "excite" us, while reason has the steadfast stability of being "calm," "dispassionate," "distant," "impartial," "aloof," "impersonal," "detached," and "cold."

3

—————

LANGUAGE AND SPEECH

The tongue of man is a twisty thing.
—HOMER*

WHILE WE USE facial expression and gesture to communicate simple emotions and thoughts, our dominant medium for transmitting information is language. Our survival, as social animals, depends on our ability to speak and understand and the power of modern culture comes from the invention of writing, some five thousand years ago. The major civilizations in the world today were all built upon the benefits of writing and reading. Libraries, with permanent records of language, are the repositories of human knowledge.

The evolutionary development of language must have been driven by the advantage of improved transfer of information, yet the role of language goes beyond communication. Language is intimately linked to the process of reasoning, but what, exactly, is the relationship between language and reason? We normally use language to reason, but are the two inextricably bound together? Can we reason without language? Can we even think without language? Great intel-

———————

* The *Iliad*, c. 800 B.C.

lects have struggled with these questions over the centuries. Claude Lévi-Strauss was pessimistic about finding answers: "Language is a form of human reason, and has its reasons which are unknown to man."* When we compare language and reason, we find that they have many features in common. Each is a biological product, apparently developed to promote better social organization and cooperation. Each can help us to reach our goals, but neither can determine what our goals will be.

In order to reason, we must think. Before moving to the specific issue of whether reasoning can occur without language, we should ask the more general question: Can we think without language? Great thinkers have thought differently about thinking. Gilbert Ryle considered that "much of our ordinary thinking is conducted in internal monologue or silent soliloquy, usually accompanied by an internal cinematograph-show of visual imagery" (1). Samuel Johnson said, "Language is the dress of thought" (2). In more technical, philosophical terms, Colin McGinn asks the pivotal question: "Is language merely the contingent manifestation of thought, required only for the communication of thoughts to others, or should we say that language is the stuff of thought, its necessary vehicle?" (3). He responds to this question by affirming its difficulty, and takes the position that as yet we have no satisfactory answer. J. J. Jenkins has offered three hypotheses: (1) thought is dependent on language; (2) thought is language; and (3) language is dependent on thought (4). He presents evidence to support each of these postulates, accepts much of the evidence as true, and concludes that the correct answer is "all of the above."

Others have taken a clearer position on one side or the other. Kant held that thinking was "talking with oneself." Max Müller went

---

* This quotation comes from chapter 9 of *The Savage Mind* (London: Weidenfeld and Nicolson, 1966). It is a play on Pascal's "The heart has its reasons which reason knows nothing of," from *Pensées* (1670).

further; he wrote a book titled *The Science of Thought: No Reason without Language; No Language without Reason.* This included a section headed "Language and Thought Inseparable," where he affirmed, "What we have been in the habit of calling thought is but the reverse of a coin of which the adverse is an articulate sound, while the current coin is one and indivisible, neither thought nor sound, but word" *(5)*. Hannah Arendt was equally forthright and sure of herself: "No speechless thought can exist" *(6)*. Ludwig Wittgenstein said the same: "The limits of my language means the limits of my world" *(7)*.

Other distinguished thinkers have taken the opposite position, arguing that language can be separated from thought. John Locke considered that words sometimes expressed thoughts, but at other times they were inadequate. Jean Piaget showed that mental processing in infants occurs before language; he therefore concluded that thought and language are distinct attributes. Frank Benson develops a similar argument, leading to the conclusion that language and thought are separable functions of the brain *(8)*. The issue has been the subject of such intense debate that we should pause to examine some of the neurological observations that indicate language is separate from thought—and reason. The essence of the argument is that neurological disorders can disconnect language from thought and reason.

LANGUAGE IS USUALLY expressed through speech, but disorders of speech occur without affecting language. Weakness of speech can be caused by disease that reduces the strength of the muscles of the tongue, pharynx, or larynx. If regions of the brain concerned with the coordination of muscles are damaged, the voice becomes slurred. To explore language, as opposed to speech, we have to examine the effects of neurological diseases that disrupt our grasp of the significance of words or our choice and combination of words into phrases and sentences. Neurological problems with language are so common that we have plenty of evidence. Selective disruption of language is of two major types: one where receiving and interpreting language is lost (receptive aphasia), and the other where the assembly and expression of language is lost (expressive aphasia). Often there is a mixture of receptive and expressive aphasia.

Recently there has been interest in subdividing aphasia according to whether the disturbance primarily involves vocabulary or grammar. Steven Pinker gives an example of aphasia in which the patient had consistent difficulty with vocabulary (9). When he was shown objects and asked to name them, he came up with related words or distortions of the correct words. The object shown is given first:

table: "chair"
elbow: "knee"
clip: "plick"
butter: "tubber"
ceiling: "leasing"
ankle: "ankley, no mankle, no kankle"
comb: "close, saw it, cit it, cut, the comb, the came"
paper: "piece of handkerchief, pauper, hand pepper, piece of hand paper"
fork: "tonsil, teller, tongue, fung"

Pinker gives another example in which grammar was lost (9). The patient was trying to describe his job in a paper mill: "Lower Falls . . . Maine . . . Paper. Four hundred tons a day! And ah . . . sulphur machines, and ah . . . wood . . . Two weeks and eight hours. Eight hours . . . no! Twelve hours, fifteen hours . . . workin . . . workin . . . workin! Yes, and ah . . . sulphur. Sulphur and . . . Ah wood. Ah . . . handlin! And ah sick, four years ago."

In his essay "The Man Who Mistook His Wife for a Hat," Oliver Sacks describes how he diagnosed a patient with an isolated difficulty in recognizing certain objects visually (10). Dr. Sacks was finishing the neurological examination.

I had taken off his left shoe and scratched the sole of his left foot with a key—a frivolous-seeming but essential test of a reflex—and then, excusing myself to screw my ophthalmoscope together, left him to put on the shoe himself. To my surprise, a minute later, he had not done this.

"Can I help?" I asked.

"Help what? Help whom?"

"Help you put on your shoe."

"Ach," he said, "I had forgotten the shoe," adding, *sotto voce*, "The shoe? The shoe?" He seemed baffled.

"Your shoe," I repeated. "Perhaps you'd put it on."

He continued to look downwards, though not at the shoe, with an intense but misplaced concentration. Finally his gaze settled on his foot: "That is my shoe, yes?"

Did I mis-hear? Did he mis-see?

"My eyes," he explained, and put a hand on his foot. "*This* is my shoe, no?"

"No, it is not. That is your foot. *There* is your shoe."

"Ah, I thought that was my foot."

In a subsequent meeting with the same patient, Dr. Sacks recounts a further conversation.

I tried one final test. It was still a cold day, in early spring, and I had thrown my coat and gloves on the sofa.

"What is this?" I asked, holding up a glove.

"May I examine it?" he asked, and, taking it from me, he proceeded to examine it. . . .

"A continuous surface," he announced at last, "infolded on itself. It appears to have"—he hesitated—"five outpouchings, if this is the word."

"Yes," I said cautiously. "You have given me a description. Now tell me what it is."

"A container of some sort?"

"Yes," I said, "and what would it contain?"

"It would contain its contents!"

The patient was capable of graphic powers of description, and he evidently had unusual skill in evasion, yet he could not recognize a glove. This kind of difficulty with handling concepts is termed agnosia.

In receptive aphasia, disease affects Wernicke's area, a small sector of the left cerebral hemisphere adjacent to the portion of the brain concerned with hearing. Another part of the brain is vital for the understanding of written language; this zone is close to the region that receives visual information. Expressive aphasia arises from a lesion in Broca's area, near the sector responsible for the execution of movement of the larynx, pharynx, and tongue; similarly, a lesion near the part of the brain concerned with movement of the right hand (in right-handed people) can result in difficulty with writing.

While lesions causing the various forms of aphasia are situated close to areas of the brain responsible for other functions related to language, these related functions are spared in aphasia. Thus in receptive aphasia there is an inability to understand spoken or written language without any defect in hearing or vision. In expressive aphasia there is a problem with speaking or writing without any disturbance in the strength or coordination of muscle contraction in the vocal tract or the hand.

The right and left cerebral hemispheres are connected by bridges of crossing nerve fibers called the corpus callosum. In certain forms of epilepsy, surgeons cut these bridges to impede the spread of abnormal electrical discharges across the brain. Studies on patients who have undergone this procedure have confirmed our ideas about the localization of language. If pictures of objects are presented, selectively, to the left side of the field of vision, they are detected in the right cerebral hemisphere. A right-handed patient, whose bridging pathways have been cut, cannot say what the pictures represent, because the right hemisphere cannot communicate adequately with the areas responsible for language in the left hemisphere. The patient, however, can choose the object correctly from an assortment of pictures. Conversely, pictures of objects presented in the right visual field are detected in the left cerebral hemisphere. Now the patient has no difficulty in naming the objects depicted, because information can readily be transferred to the area controlling speech within the same (left) hemisphere.

ONE ODDITY NEEDS to be mentioned for the sake of completeness: the curious phenomenon of speaking in tongues, otherwise known as glossolalia. This consists of outbursts of unintelligible utterances from individuals who seem to be undergoing an ecstatic, mystical experience. Over the centuries, heated debates have taken place concerning the nature of these episodes. Ecclesiastics have played a prominent role in the discussions because glossolalia usually takes place in a religious setting. With modern recording techniques it has become possible to transcribe and analyze glossolalia. As a result we can now say quite firmly that glossolalia does not represent the sudden spontaneous acquisition of unfamiliar languages. Glossolalia is gibberish *(9)*.

THE STRUCTURE OF LANGUAGE
Language and reason are tightly bound together even if, under certain circumstances, they can be split apart by neurological disease. We communicate reason through language. We even build major instruments of reason—arguments—out of language. We should, therefore, see what the science of linguistics has revealed about the nature of language, for this information may give us some inkling of what to expect if a similar analysis is applied to reason

Language is a highly organized system of symbols used for the transmission of information; the symbols can be stored, retrieved, and manipulated by the brain. To retain the flexibility necessary for a rich and varied language, the smallest components, when isolated, must not have any intrinsic significance. Sounds such as *p*, *a*, and *t* are called phonemes. Meaning comes from the way individual phonemes are combined, for example in *pat* or *tap*. This principle is evident in other biological settings. For instance in molecular biology, the messages that control cell structure and function are determined by the order in which coding units—base pairs—are combined within DNA. Just as each phoneme, by itself, has no meaning, so each base pair, by itself, has no meaning. Everything depends upon the order in which the phonemes or base pairs are assembled. This design is extremely efficient for both language and molecular biology, because the same coding units are used again and again in different combinations. Molecular biology has adopted terminology that resonates with linguistics; people studying DNA structure and function talk about translation, transcription, reading frames, and libraries.

The simplest combinations of phonemes that acquire meaning are termed morphemes, which comprise words and word elements such as affixes and inflectional endings (*re-, -un, -ation, -ed, -s*). While morphemes have some meaning when they stand alone, their meaning is extended greatly by the formation of phrases and sentences through the application of grammar. All of these features allow language an endless variety of sentences, and a limitless capacity for novelty by adding sentence after sentence.

Linguistic scientists infer the existence of a layer of organization that underlies the vocabulary and grammar of language, a semantic system responsible for interpreting meaning. It functions smoothly, without our direct awareness of its existence, yet its mechanism of operation is poorly understood. The meaning denoted by a word or sentence is one of the most important elements of the infrastructure of reason, yet even the concept of "meaning" is obscure. The philosopher and the linguist approach the analysis of this problem quite differently. Zeno Vendler has summarized these divergent approaches:

"Whereas the philosopher asks the question 'What is meaning?,' the typical questions the linguist is likely to ask include 'How is the meaning of words encoded in a language?,' 'How is this meaning to be determined?,' 'What are the laws governing change of meaning?' and 'How can the meaning of a word be given, expressed or defined?' " (*11*). These ideas are helpful, but they do not adequately convey the problem of understanding meaning. Colin McGinn concludes that the meaning of meaning is beyond our grasp: "What is the meaning of a word? Words have it in common that they contribute toward the meaning of sentences, but little else can be said to unify them; in the same way it may be that concepts share no significant features beyond the fact that they contribute to the content of thoughts" (*3*).

Philosophers have always recognized the importance of language. In recent years the school of linguistic analysis has argued that many philosophical problems are caused by confused or misused language, so careful and precise reformulation of language should solve or dispel the problem. This stance is not very satisfactory for a biologist seeking some understanding of language as a product of evolution. A more helpful approach is to consider meaning as the link that ties a sentence with a thought—a sentence "means" the thought that it generates. A thought will generate a sentence, and the sentence will reflect back the thought. The idea that "meaning" is the link between words and thoughts offers a useful working definition, even if it is an oversimplification.

The next obvious questions are how does a thought acquire meaning, and what is the meaning of a thought? Under philosophical scrutiny the relationship between thoughts and words, and the definition of meaning, are far from straightforward. Numerical definitions allow us to be sure of statements such as "two plus three equals five," but beyond the special, internally consistent system of mathematics, what do words such as "true" and "false" mean? Statements about the physical world can surely be true (such as "Ice covers the

South Pole") or false (such as "Canada is located on the Equator"). "True" and "false" have a ring of universal, constant, and eternal meaning, but this impression of certainty is misleading. Hilary Putnam persuasively argues that our culture determines what is generally agreed to be true and false, because circumstances and attitudes establish the criteria for the acceptability of a statement *(12)*. Three thousand years ago, most well-informed people concluded, from the evidence and arguments then available, that the world was flat. Gradually, new observations and new ways of thinking forced a reassessment, pioneered by the Pythagoreans, so now we are prepared to assert that the world is round. What we accept differs according to time and place, so if we say that a statement about the world is true or false we are not speaking with certainty, and we have no absolute terms of reference. We are talking about probabilities, and the best that we can hope for is a position comparable to that demanded of testimony in criminal law—a statement may be regarded as "true" if it can be affirmed "beyond reasonable doubt" according to the criteria operating at a particular time, in a particular place, for a particular culture.

Benjamin Lee Whorf has proposed a theory of linguistic relativity. He emphasizes the variation in conceptual content of different languages and suggests that special cultural features are responsible: "We are thus introduced to a new principle of relativity, which holds that all observers are not led by the same physical evidence to the same picture of the universe, unless their linguistic backgrounds are similar" *(13)*. While the interpretation of physical events may be influenced by language, the consistency of judgments poses an even greater problem. Is it true that "Beer tastes better than wine," or that "Bach was a greater composer than Beethoven"? When applied to these assertions, the notions of "true" and "false" seem hopeless.

While some statements do not lend themselves to being described as "true"—in the sense of "beyond reasonable doubt"—we should not infer that "true" and "false" have no meaning. We simply have to be cautious how we interpret words, and recognize their limitations. "Truth" has proved a very useful notion in scientific exploration of the nature of the world, and the fact that it only means "beyond reasonable doubt" does not detract from its value.

IN ADDITION TO all these difficulties, the context of words can have a profound effect on meaning. For example, "Animals are our friends" has two morphemes, "our" and "are," that sound similar. We interpret these morphemes correctly because of the linkage to less ambiguous morphemes, "animals" and "friends." But "animals" and "friends" have their own nuances. "Animals are our friends" has quite a different meaning from "Our friends are animals." By changing the order of words, we alter meaning. From these examples we can see that words vary in their dependence on context for meaning; indeed, words can even be classified according to how "context-bound" or "context-free" they may be. The interpretation of entire sentences can also be influenced by context. For example, "I will see you tomorrow" may be a prediction, a promise, or a warning, depending on who says it to whom, and in what circumstances. Literary works thrive on these diversities. Poetry would be impossible without the rich assortment of feelings evoked by a few carefully assembled words. The force of poetry depends upon the words selected and how they are put together. We marvel at the result because we do not expect such a direct and powerful impact on our feelings.

While language has all of these sophisticated facets, it is also a rough-and-ready biological product that gives us an evolutionary advantage because it improves social cooperation and it allows the development of more versatile patterns of behavior. Students of language explore how it has been polished and refined to take on highly specialized and demanding roles. Many languages have met the needs of the arts and sciences without difficulty, but none has been able to withstand the joustings of philosophy. Skillful intellectual probing reveals many inconsistencies and incoherences in the use of language. As an example, what does it mean when a message on one side of a piece of paper says the information on the other side is false, and the message on the other side says the same thing? Perhaps philosophers are pushing language beyond its operational capacity so that it breaks down just like any other instrument subjected to too much stress.

WHERE DO GRAMMAR and vocabulary come from? Many linguistic scientists have emphasized that all the five thousand current languages (and all previous languages that we know about) are made up of an infinite number of sentences, each assembled according to certain principles (9). They argue that these shared "language universals" have evolved by natural selection. According to this theory, children have an inborn awareness of how to form sentences, but they learn their vocabulary from their environment.* Several sources of evidence can be cited to support this view: (1) similar principles for constructing sentences underlie all languages; (2) children assemble sentences before they could have received sufficient exposure to language to have learned grammar; and (3) there are neurological disorders of language in which the ability to build sentences is selectively lost or preserved, suggesting that there is a distinct functional module in the brain. We have already seen how grammar may be primarily lost in certain forms of aphasia, and as another example of disease targeting grammar, there are families with a hereditary condition called "specific language impairment" (14). Affected individuals are normal apart from having very slow, labored language with numerous grammatical mistakes. An argument for a specific grammar "module" can also be developed from observations on patients who suffer from symptoms that are the opposite of expressive aphasia and specific lan-

---

* In 1959 a heated debate developed between B. F. Skinner and N. Chomsky; it was destined to have a far-reaching impact on the study of language. Skinner argued that children learn all their language from their parents, while Chomsky held that they have an inborn ability to create sentences, so they only have to learn a vocabulary.

guage impairment—neurological disorders exist where language is selectively preserved in spite of extensive disruption of other mental processes. Williams' syndrome, a rare genetic form of mental retardation, falls into this category *(15)*.

In contrast to grammar, vocabulary must be learned because words differ from one language to another. Words pass from generation to generation and become key features of the local social environment. But as cultures meld in the modern world, there are many obvious advantages to be gained from seeking a shared language. The difficulty in achieving this goal reflects the power of language as a symbol of separate cultural identity and a tool for ideological persuasion. We have only to look at the Francophone-Anglophone conflicts in Canada, where the provincial government of Quebec employs "language police," also known as "tongue troopers," to find and suppress unapproved languages.

MATHEMATICS

Variations in the interpretation of words can be exploited by poets, but for some tasks a more strictly defined form of communication is required—the discipline of mathematics. Perhaps mathematics is the quintessential expression of reason. Mathematics was built with new symbols, new conventions, and new rules. The essential features of mathematics are (1) brevity, (2) consistency, and (3) precise abstraction. Numbers and mathematical operations allow no ambiguity. The concepts of "meaning," "truth," and "certainty" are easier to handle in this setting. Even when we address questions that are inherently uncertain, mathematics can quantify the levels of probability.

Carl Boyer makes an educated guess about how our early hominid ancestors began to count:

At first the primitive notions of number, magnitude and form may have been related to contrasts rather than likenesses—the difference between one wolf and many, the inequality in size between a minnow and a whale, the unlikeness of the roundness of the moon and the straightness of a pine tree. Gradually there must have arisen, out of the welter of chaotic experiences, the realization that there are samenesses; and from this awareness of the similarities between number and form both science and mathematics were born *(16)*.

Different single objects would be recognized as dissimilar to each other, but in a sense similar just because each is "single." Pairs would be identified as another category; hands, feet, eyes, and ears would be placed in this class. Then the concept of "two hands" would be broken down into "two" and "hands." The benefit of these discoveries would become apparent because the abstraction of "two" had useful applications beyond its attribution to hands, feet, eyes, and ears. Two would add valuable additional information when applied to items that did not normally come in pairs, such as fish or axes. As Boyer writes, "The recognition of an abstract property that certain groups hold in common, and which we call number, represents a long step toward modern mathematics." It is also a long step forward for reason.

Specific symbols for numbers, such as notches, were employed before written words, and a counting repetition base of five preceded the base of ten. For example, the bone of a young wolf, found in central Europe, has fifty-five notches cut deeply into it *(16)*. The bone predates the invention of writing and the notches on the bone are organized into two large groups, of twenty-five and thirty, respectively. The cluster of twenty-five is arrayed in five sets of five, and the batch of thirty is made up from six sets of five, so presumably our forebears counted on the fingers of one hand before they began to use both. Seven would be five and two, nine would be five and four, and so on. Repeated counting on the fingers of one hand was easy for small numbers, but in due course the need for large numbers must have led early mathematicians to abandon the base of five, and move on to using all ten fingers. The long association between fingers and numbers is implicit in the use of the word "digit." While we have a plausible explanation for the widespread adoption of the decimal system, some cultures have explored other numerical bases such as binary and ternary systems. Higher bases have also been used; for example, several American Indian tribes adopted a base of twenty.

THE CONCEPTS OF simple arithmetic seem to be genetically determined, "hard wired" in the brain of *Homo sapiens* like the principles of grammar. Infants have a rudimentary capacity for handling numbers before they can even talk.* As for language, symbols initially employed for communication acquired much more power when they were used within a system of rules that enabled them to be combined for special purposes, such as solving problems. With increasing experience in mathematics, people found that numbers could be grouped in different ways, and a further set of symbols was introduced to represent a range of numbers with a particular role. The most readily available symbols were letters, so algebra was born and general relationships became available for mathematical operations. For example, $y = mx + c$ summarizes all possible options for a straight line located between two axes in two dimensions, where x and y are variable numbers that fit onto the line, m represents a constant that defines the slope, and c is another constant that defines where the line cuts an axis.

---

* Karen Wynn has reported early mathematical ability in her paper "Addition and subtraction by human infants," *Nature*, 358, 749–750, 1992. Other evidence supports the existence of a separate module of brain function for manipulating numbers; defective calculation (acalculia) can be the major feature of developmental disorders of early life, and in acquired neurological disease of late life such as stroke—see Todd E. Feinberg and Martha J. Farah, *Behavioral Neurology and Neuropsychology*, New York: McGraw-Hill, 1997.

The rules for manipulating numbers in mathematics are obviously different from the conventions of grammar in language, and each system has its uses. In general, mathematics provides solutions to problems, and language provides descriptions, but there is overlap. Sometimes a mathematical formula can describe the features of a curve better than words, and words can be used to solve problems when they are employed in a logical construct. A specific verbal argument may run: "All men are mortal; Socrates is a man; therefore Socrates is mortal." The corresponding but more general mathematical proposition will run: "If a and b are two classes and a is contained in b, then x is in a implies that x is in b" (17).

Mathematics became essential for the operation of large communities, where resources have to be allocated and a monetary system is necessary. Early civilizations used mathematics for architecture and engineering. Later, mathematics became central to almost every field of science and technology. It is usually associated with the physical sciences, but its achievements in the biological sciences have also been spectacular. While mathematics has solved many old problems, it has also generated many new ones. Twentieth-century mathematical concepts such as the expanding universe, Heisenberg's uncertainty principle, and Einstein's general law of relativity are difficult for us to handle intellectually; we are hampered by our traditional, self-contained and biologically useful mental framework of space, time and causation.

EVOLUTIONARY ORIGINS OF LANGUAGE

Bees proclaim where they have found nectar by dancing at an angle from vertical that corresponds to the angle between the sun and the direction of the source of food. The intensity of abdominal waving in the dance gives an indication of distance. In spite of their relatively simple nervous system, bees can compute adjustments to the information that they communicate. If the sky is clouded over when they return from foraging, they will make appropriate alterations in the angle of their dance to adjust for the fact that the position of the sun—which they cannot see—is changing with the passage of time. Different communities of bees develop their own minor variations in dance, called "dialects" by those who equate insect communication with human language. Other species can also convey precise information—vervet monkeys have specific alarm calls for dangers that require different communal responses. If an eagle is threatening from above, the warning tells the troop to climb down from the trees. When a predator is sighted on the ground, the call sends the troop scurrying up the trees.

HUMAN LANGUAGE PROVIDES an enormous increase in the quantity of information available for transfer, helping intellectual activity in general and the faculty of reason in particular. We take for granted our special ability to describe abstractions—such as "tomorrow"— yet such abstractions are only made possible by language. Another invaluable feature of human language is the ability to ask questions. In their thoughtful analysis of the differences between communication in animals and humans, David and Ann Premack stress the importance of questioning. Animals transfer information through codes that have "a correlation between certain acts in the sender and certain items in the world. . . . A code becomes a language if the user can be questioned about the code" *(18)*.

Pinker argues that grammar is the unique feature of human language that sets it apart from communication among animals:

Language is obviously as different from other animals' communication systems as the elephant's trunk is different from other animals' nostrils. Nonhuman communication systems are based on one of three designs: a finite repertory of calls (one for warnings of predators, one for claims to territory, and so on), a continuous analog signal that registers the magnitude of some state (the livelier the dance of the bee, the richer the food source that it is telling its hive mates about), or a series of random variations on a theme (a birdsong repeated with a new twist each time). . . . As we have seen, human language has a very different design. The discrete combinatorial system called "grammar" makes human language infinite (there is no limit to the number of complex words or sentences in a language), digital (this infinity is achieved by rearranging discrete elements in particular orders and combinations, not by varying some signal along a continuum like the mercury in a thermometer), and compositional (each of the infinite combinations has a different meaning predictable from the meanings of its parts and the rules and principles arranging them) *(9)*.

Speech is crucial for normal human language. It requires the physiological capacity to utter a vast range of different sounds in rapid succession, and each sound must be readily distinguishable. There is a need for (1) suitable vocal anatomy, (2) appropriate pathways in the brain, and (3) adequate auditory acuity. The assignment of meaning to selected sounds will depend, at least in part, on the ease with which they can be uttered and heard. A sentence of twelve words, comprising some fifty phonemes, can be communicated in two seconds—a rate of twenty-five items of auditory information per second. In contrast, outside the context of speech, comparable auditory stimuli presented as "noise" can only be distinguished at a rate of nine items per second. Sounds are recognized more readily because they are presented in familiar linguistic combinations, and also because language is constructed from phonemes that are easy to distinguish from each other.

The biological machinery for human speech is unique. To achieve an adequate range of sounds, evolution had to move the larynx down from the back of the mouth, where it is situated in monkeys and apes, to the neck, where it is situated in *Homo sapiens.* The descent of the larynx into the neck enabled specialized anatomical changes to take place above the larynx; these structural developments allowed a far more extensive range of phonation. But the restructuring of the vocal tract could only be achieved at a cost. It was no longer possible to breathe and eat (or drink) simultaneously; choking, a potentially lethal malfunction, became more frequent. These difficulties may explain why speech has only evolved in hominids who had the cerebral capacity for language of such power that it would give an overwhelming biological advantage.

Developments in vocal anatomy were accompanied by changes in the brain to support the operation of language, but we have no idea when all these developments occurred. Our hominid ancestors seemed to have possessed language, but we know nothing of the form it may have taken, or how it would have been used. We are only just beginning to understand the complexity of the neural organization required. First, we need neural mechanisms to link language with thought, and to apply grammatical principles. Then we need to transform sentences into speech. An operational model for speech entails a stage of "phonological encoding," where we assemble the "phonetic plans" and store them in an "articulatory buffer." When we are ready to speak, the plans are retrieved from the buffer and "unpacked" for translation into motor commands. Patterns of nerve impulses must then be transmitted to the muscles of the larynx, pharynx, tongue, lips, diaphragm, and chest wall. The neurophysiology of speech and language is far from simple.

We can only speculate on how these building blocks of speech and language evolved. The problem is that natural selection generally operates upon small incremental changes in structure and function. Yet the emergence of hominids was, in biological dimensions of time, sudden and spectacular. Pinker poses two analogies that illustrate how language could have arisen as either an improvement in the existing capacity to communicate in other primates, or the development and reorganization of brain circuits previously serving other functions.

An example of a new module is the eye, which has arisen de novo some forty separate times in animal evolution. It can begin as an eyeless organism with a patch of skin whose cells are sensitive to light. The patch can deepen into a pit, cinch up into a sphere with a hole in front, grow a translucent cover over the hole, and so on, each step allowing the owner to detect events a bit better (9).

Pinker then cites the elephant's trunk to illustrate the reorganization of existing structures to produce something new. The elephant's trunk "is a brand-new organ, but homologies suggest that it evolved from a fusion of the nostrils and some of the upper lip muscles of the extinct elephant-hyrax common ancestor, followed by radical complications and refinements."

ACQUISITION OF LANGUAGE

Having considered how our ancient forebears evolved speech and language, we will now see how each new individual obtains it. Children learn language much more readily than adults. Comprehension of language is gained before words are spoken; the capacity for decoding the symbols of language precedes the ability to encode them. Language advances as syllables are uttered; single morphemes may be expressed—usually nouns—around the age of twelve to twenty months. At this early stage the infant is learning a vast range of skills. A chaotic abundance of new diverse experiences must be integrated into an organized mental "database."

Words that are acquired early have a wider meaning for children than for adults. For example "ball" may mean "Where is the ball?" or "There is the ball!" or "Give me the ball." Then combinations are made between two words. Some linguists have designated certain words as "pivotal" because they are frequently joined with other words to form phrases. For example, "more" is pivotal in combination with "milk," "cake," or "play." From one or two words at the age of one year, the child's vocabulary rises to 200 by the age of two years. Grammar brings a big improvement to communication around the age of three and a half years. An average English-speaking adult has a vocabulary of about 50,000 words, but there is considerable variation. University graduates have a repertoire of some 80,000 words.

The development of language coincides, to some extent, with the major milestones of general motor function. The extremely precise and highly organized integration of muscle contraction required for speech, involving the lips, tongue, pharynx, larynx, and diaphragm, precedes comparable control of fine movements of the fingers.

Language is managed by the left side of the brain in 95 percent of right-handed individuals, but only 70 percent of those who are left-handed. The association between language and handedness suggests that communication and the ability to use tools evolved together. Although individual animals, particularly apes, may have a minor preference for using one or the other hand, the phenomenon of handedness—consistently greater use and skill on one side—is uniquely developed in *Homo sapiens*. We do not know why language is concentrated in one cerebral hemisphere. Michael Gazzaniga suggests that "one and only one linguistically dominant hemisphere is in charge of language because of the nature of speech" *(19)*. The integrated control of very rapid movements of the muscles of speech on both sides of the larynx, Gazzaniga argues, could not be directed from both of the cerebral hemispheres because transfer of information between the hemispheres, to coordinate the laryngeal muscles, would simply take too long.

What we know about language is far less than what we do not know. The neurophysiology of language includes highly intricate processing of information in the brain, exquisitely precise muscle control in the larynx, and finely tuned auditory acuity in the ear. The power of language is spectacular, but most of the time we are not aware of it.

Reason, like language, has its own rules, and we learn to use these rules at a young age. We know that reason and language developed together and are mutually dependent, though we also know that they came from separate modules in the brain. As yet, we lack access to a level of analysis that can specify the precise nature of their interaction.

4

SOCIAL BEHAVIOR

*Social primates are required, by the very nature of the system
they create and maintain, to be calculating beings; they must
be able to calculate the consequences of their own behavior,
to calculate the likely behavior of others. . . . "social skill"
goes hand in hand with intellect.*

—NICHOLAS HUMPHREY*

Two thousand years ago Seneca, the Roman orator, philosopher, statesman, and millionaire, wrote that "man is a reasoning animal" *(1)*. In 1677 Spinoza declared that "man is a social animal" *(2)*. We reason and we socialize, building upon patterns of behavior that are widespread in the animal kingdom, but these are not two independent activities. The links between reasoning and socializing can be found by looking at the impact of reason on society and the inpact of society on reason.

We have achieved our present fund of human knowledge through contributions made by billions of people over hundreds of thousands of years. Early nomadic communities obtained and passed on practical information for survival, such as the design of tools, con-

---

* "The Social Function of Intellect," in *Growing Points in Ethology,* ed. B.P.G. Bateson and R. A. Hinde (Cambridge: Cambridge University Press, 1976).

trolled use of fire, methods of hunting, and local geography. Later cultures, including the ancient Greeks, prized reason as the supreme gift of the gods, and they used it to accelerate the extension of knowledge. They promoted it through its social expression in the form of discussion and debate. Reason has been encouraged, tolerated, manipulated, and suppressed in countless social settings from families to religions, from universities to political regimes. Just as society has shaped reason, so reason has shaped society. Advances such as the invention of steel, gunpowder, and the internal combustion engine were the direct result of applied reason, and these technologies have revolutionized the cultures that created them.

Is this tight linkage between reason and society simply coincidental, or is there a causal relationship binding them together? A challenging recent idea envisions intelligence, with its capacity for reason, as an evolutionary product gained because of the survival advantages that early hominids achieved from living in social groups. More specifically, intelligence was required for the complexity of social skills necessary for successful communal life. This view has been argued eloquently by Nicholas Humphrey. He starts with a definition of intelligence in animals: "An animal displays intelligence when he modifies his behavior on the basis of valid inference from evidence" (3). He then points out a paradox: studies on monkeys and apes indicate the presence of far more intelligence than they seem to need for their normal environment. Why should this "spare capacity" have evolved? Humphrey suggests that the extra intelligence, beyond that obviously required for the necessities of daily living, is essential for maintaining satisfactory social interaction.

With this background, an examination of how animals live together allows us to see the influence of communities on the development of reason, and the influence of reason on the development of communities.

SOCIAL BEHAVIOR OF ANIMALS

Some features of social life among animals come from instinct, while others, particularly in the monkeys and apes, can only be explained by a capacity for reasoning. On an evolutionary time scale, instinct is old and reason is new. Insects, devoid of reason, operate in vast, highly ordered populations that can dominate their environment. Isolated from the colony, lone members have a negligible chance of survival. The power generated by large numbers of ants is vividly illustrated by Edward O. Wilson:

> Viewed a few meters away, a driver-ant raiding column seems a living thing, a giant pseudopodium reaching out to engulf its prey. The victims are snared with hook shaped jaws, stung to death, and carried to the bivouac, a labyrinth of underground tunnels and chambers housing the queen and immature forms. Each expeditionary force comprises several million workers who flow out of this retreat. The hungry legions expanding from the bivouac are like an expanding sheet that lengthens into a treelike formation. The trunk grows from the nest, the crown expands as an advancing front, and numerous branches pour back and forth between the two. . . . The front, advancing at 20 meters an hour, blankets all the ground and low vegetation in its path. The columns expand into it like a river entering a delta, where the workers race back and forth in a feeding frenzy, consuming most of the insects, spiders, and other invertebrates in their path, attacking snakes and larger animals unable to move away *(4)*.

This shows how large social groups get power, without individuals exercising the faculty of reason. Here we have a paradox: when it comes to huge social enterprises, reason may be unnecessary—indeed it may even be an encumbrance! If ants could reason, would they continue to sacrifice themselves in enormous numbers? If Wilson's evocative account of ants is transferred from its biological setting, it could be an allegory of the massed infantry formations in the First World War, where human armies adopted the tactics of ants in similar large-scale "steamroller" maneuvers. The Battle of the Somme was sustained by cultural imperatives that defied reason. After the initial failure of the British attack, the infantry on each side did not allow reasoning to interfere with orders. Large numbers of individuals can be transformed into a disciplined biological machine when their efforts are united for a common purpose. Reason is not a useful instrument for evolutionary survival in this setting. The social behavior of individual ants lacks flexibility, yet the ant family is very successful in biological terms and its success springs from rigid social organization. There are about a million times as many ants in the world as humans. Expressing this statistic in another way, the total mass of ants is approximately equal to that of humans (an average human weighs about the same as a million ants).

MAMMALS, OTHER THAN modern *Homo sapiens,* operate in much smaller communal organizations than ants. For hunting, packs of hyenas adjust their size and tactics according to their prey. Cooperative groups of hyenas attack herds of zebras because counterattacks can be anticipated; but if the prey is a single antelope, one hyena will take responsibility for the chase. Defense is also handled more efficiently by groups of animals acting cooperatively—potential victims band together and take specialized responsibilities. Male baboons and patas monkeys adopt the role of sentinels while their troops forage.

The internal structure of groups of primates varies between species. In macaques and baboons there is a clear ranking of dominant adult males in each troop; the order is determined by their success in fighting. Here dominance may arise from overt tests of strength or there may be obvious manifestations of physical power, such as size, that make contests unnecessary. Subordinate animals

accept a lower priority in the competition for food and mates because the alternatives are worse: solitary survival is difficult for an
animal that normally lives in a group and emigration to another
community is unlikely to improve social status when newcomers
normally go to the bottom of the heap.

In addition to this linear dominance, coalitions form. Robert
Trivers witnessed a scene in Africa that illustrates how such alliances
can take place among baboons *(5)*. He named the players Arthur,
Anne, and Carl. Arthur is the dominant head of the troop; Anne is
highly receptive to amorous advances; Carl is very interested in Anne
and also happens to be a member of a subversive group undermin-
ing the authority of Arthur. Trivers describes how Carl's alliance
frustrates Arthur's attempts to exercise his *droit du seigneur* with
Anne. Solidarity within the coalition allows Carl to have his way with
Anne in spite of Arthur's jealousy. On three occasions over the course
of a day, the alliance intervenes to prevent Arthur from interrupting
Carl's sexual intercourse with Anne.

> Carl would mount Anne and start thrusting, at which point
> Arthur would rush in to try to break up the copulation, and one of
> the other males—whichever happened to be closer—would re-
> spond by interposing himself between Carl and the onrushing
> Arthur. If this was successful, then Carl would be able to bring his
> copulation to completion, and the two males would join together
> and chase Arthur off.

Trivers introduces the social concept of jealousy, showing that baboons have a notion of possession for mates, just as they have for territory.

MUTUAL HELP—RECIPROCAL altruism—is widely practiced by
social animals. Reciprocal altruism is behavior of one animal that
helps another, with the expectation that the help will be repaid so
that the benefits for all exceed the costs. Over a reasonable period of

time there will be a net gain for everyone involved. The only problem is the temptation to cheat by receiving help without returning it. In a stable group, breaking the code of behavior will mean that sooner or later the cheat will be denied help. The key to the success of reciprocal altruism, as a tool for sustaining social cooperation, is the ability to take action against those who fail to reciprocate.

For spider monkeys and chimpanzees, strong emotional ties develop between females and their offspring so that closely knit matrilines stabilize the framework holding the community together. Females spend their time being pregnant, lactating, or caring for their offspring. Males commit less time to social activity; they mate, groom, and roam. Male mobility provides a mechanism for interbreeding with different populations of the same species. Such mixing of the genes is a healthy feature from the viewpoint of evolution.

As in human societies, family relationships are important in monkeys and apes, so when fights occur animals tend to help their closest relatives. If an ape fails to join in a fight involving a relative, an obvious explanation is usually apparent. Apes seem to assess the chances of winning, and when the odds are bad enough they avoid the fray. The choice of discretion over valor is, in this circumstance, likely to conserve the genes for circumspection and perhaps favor the selection of wise rather than foolish siblings.

The size of communities of monkeys and apes depends on both genetic and environmental factors—the species and the territory. The distributions of food, water, and predators are all relevant. Where the resources are plentiful and safe, groups often fractionate, but when threatened by deprivation or danger, groups coalesce. This obviously mirrors human behavior. Some species display considerable flexibility in social organization, breaking up into small, more efficient subunits for daily activities. Hamadryad baboons congregate in large troops to sleep at night, and each morning they divide into small groups that set out for the foraging grounds. Further dispersal then takes place, into yet smaller units consisting of one male with one or two females and their young. These compact teams collect food until early afternoon, when they meet together at a watering hole before returning to sleep with the combined troop.

Social interactions can be subtle and complex in communities of monkeys and apes. The most highly developed behavior appears when animals are dealing with each other individually, but their combined behavior is also impressive. Communities hold on to their territory and adopt firm policies on the extent to which their boundaries will be protected. There is naturally a cost to be paid for territoriality, and decisions seem to be determined by weighing the risks of warfare against the benefits of the resources—supplies of food and water. Defense may involve aggressive vocalization, hostile displays, and direct physical fighting. Apes and monkeys also threaten by banging anything that will make a noise, and by displaying their erect penes. Eibl-Eibesfeldt describes how sentinel vervet monkeys develop an erection if an unknown member of the same species approaches in a hostile way: "These 'guards' presenting their genitals play the part of living frontier posts. Apart from their genitals, they also make a display of a threatening face. In its origin this behavior can very likely be explained as a ritualized threat to mount" *(6)*.

IT IS HARD to escape seeing human metaphors in the social life of monkeys and apes: the dominance hierarchies with subversive conspirators, the "you scratch my back and I'll scratch yours" underlying reciprocal altruism, the wandering males and the stable females with their young, the unity of purpose in defending communal resources, the group solidarity when times are bad and its breakup when times are good. Social organization is a rational response to the demands for survival in unpredictable surroundings. The capacity for reasoning in monkeys and apes is entirely in keeping with their social structures. J. F. Wittenberger concludes that perhaps the most important questions "revolve around the ways that individuals pursue their own best interests within social contexts and how these pursuits are constrained by the activities of others" (7). The patterns of social life in monkeys and apes have inspired psychologists to coin the term "Machiavellian intelligence."

MACHIAVELLIAN INTELLIGENCE

In 1988, Richard Byrne and Andrew Whiten published a book titled *Machiavellian Intelligence: Social Expertise and the Evolution of Intellect in Monkeys, Apes, and Humans* (8). They were impressed by the way monkeys and apes employed the same duplicity that Niccolò Machiavelli described in his analysis of how princes obtain and keep political power (9). Byrne and Whiten explain that they had not, initially, intended to dwell on behavior designed to "deliberately manipulate, exploit, and deceive social companions," but "the fact is that the strong thrust in the data available at present is Machiavellian: in most cases where uses of social expertise are apparent, they are precisely what Machiavelli would have advised!"

Emil Menzel offers a vivid example of the phenomenon in chimpanzees. The major characters are Rock, a dominant male, and Belle, a clever female. In an experiment where food was hidden in a field, Belle was proficient at finding it, and once found, she shared it with most of the other chimpanzees—providing Rock was not present. When Rock appeared, Belle became

> increasingly slower in her approach to the food. The reason was not hard to detect. As soon as Belle uncovered the food, Rock raced over, kicked or bit her, and took it all. Belle accordingly stopped uncovering the food if Rock was close. She sat on it till Rock left. Rock, however, soon learned of this, and when she sat on one place for more than a few seconds, he came over, shoved her aside, searched her sitting place, and got the food. Belle next stopped going all the way. Rock, however, countered by expanding

the area of his search through the grass near where Belle had sat. Eventually, Belle sat farther and farther away, waiting until Rock looked in the opposite direction before she moved toward the food at all—and Rock, in turn, seemed to look away until Belle started to move somewhere. On some occasions, Rock started to wander off, only to wheel around suddenly precisely as Belle was about to uncover the food *(10)*.

To get the food, Belle was employing "cunning," as advised by Machiavelli. Once she had the food, however, Belle shared it in a way that Frans de Waal has called "good natured" *(11)*. Reason is a key element in both "cunning" self-interest and "good natured" cooperation.

Was subtle social interaction simply a manifestation of intelligence that had been developed for some other purpose, or was social interaction so advantageous for survival that it became the driving force for the evolution of intelligence? Was human reason fashioned from that extra bolus of intelligence that provided expertise in dealing with increasingly complicated social situations? From the evidence available, the answer must be "perhaps." In the evolution of *Homo sapiens,* mutually supportive biological advantages were gained from (1) skillful personal interactions within the community (self-interest and cooperation, both entailing reason), (2) language (to improve communication and extend reason), and (3) manual dexterity (to hunt, gather, make clothing, and build shelters). These three attributes—social skills, language, and dexterity—led to the establishment of cultures whose achievements were reflected back to enhance the three capacities that made the evolution of the cultures possible. This positive feedback has resulted in an explosion of cultural innovation.

EARLY CULTURES

In our quest to trace the origins of human reason, we must look back into the murky obscurity of prehistoric evidence to bridge the gap between apes and hominids. Prehistory is, by definition, the era that preceded written records, so our ability to reconstruct events is rudimentary. We have bones, stones, cave paintings, burial sites, and figurines to help us.

The pivotal link between apes and humans was earliest known hominid, *Australopithecus afarensis,* who appeared some 4–4.5 million years ago as the first hominid. From this time onward, our ancestors were undergoing substantial changes in brain size, body weight, and physical appearance. Preliterate cultural history really began with the appearance of *Homo habilis* some 2 million years ago. *Homo sapiens* emerged in an archaic form some 400,000 years ago, and in a modern form, 200,000 years ago.

The evolution of culture went hand in hand with biological development; they depended upon each other and they drove each other. Clifford Geertz sketched out the nature of this interrelationship:

> Culture, rather than being added on, so to speak, to a finished or virtually finished animal, was ingredient, and centrally ingredient, in the production of that animal itself. The slow, steady, almost glacial growth of culture through the Ice Age altered the balance of selection pressures for the evolving *Homo* in such a way as to play a major directive role in his evolution. The perfection of tools, the adoption of organized hunting and gathering practices, the beginnings of true family organization, the discovery of fire, and most critically, though it is as yet most difficult to trace it out in any detail, the increasing reliance upon systems of significant symbols (language, art, myth, ritual) for orientation, communication and self-control all created for man a new environment to which he was then obliged to adapt. . . . Between the cultural pattern, the body, and the brain a positive feedback system was created in which each shaped the progress of the other, a system in which the interaction among increasing tool use, the changing anatomy of the hand, and the expanding representation

of the thumb on the cortex is only one of the more graphic examples* *(12)*.

Some 97.5 percent of the era of modern *Homo sapiens* was preliterate history. It witnessed an impressive list of achievements: the appearance of religion and art; the manufacture of bows, arrows, and spears; the construction of shelters and boats; the controlled use of fire for cooking and heating; the smelting and working of metals; the production of ceramics; the discovery of agriculture; the domestication of animals; the invention of the wheel. These developments were astonishingly rapid compared with the normal time course of geological and biological change. As Geertz argued, cultures were catalysts for innovation, and the same point has been made by Stephen Jay Gould: "If I invent the first wheel, my brainchild is not condemned to oblivion by hereditary impassability (as any purely bodily improvement would be). I just teach my children, my apprentices, my social group, how to make more wheels. The point is simple yet so profound"† *(13)*.

Our knowledge of preliterate history comes from archaeology, and it is mostly the story of hominids that came before *Homo sapiens*. The importance of stone tools is reflected in the classification of preliterate history as old (Paleolithic), middle (Mesolithic), and new (Neolithic) "stone cultures." While the duration of these cultures varied from place to place, the Paleolithic age ran from about 2.5 million to some 15,000 years ago; the Mesolithic age extended from the end of the Paleolithic period to some 10,000 years ago; and the

---

* Electrical stimulation of certain areas of the brain elicits movement in different parts of the body, and damage to these areas leads to weakness in the corresponding regions. These regions of "functional localization" for movement differ in size, one of the largest being responsible for the thumb.

† As a more contemporary example of the same principle, mutation and natural selection would eventually lead to better rates of recovery from infective disease, but the cultural development of sanitation, vaccination, and antibiotics achieved the same result far more quickly.

Neolithic age ran from the end of Mesolithic time to around 5,000 years ago, when metal technologies ushered in the bronze age, and then the iron age, which in turn led into the first stages of literate history.

The Paleolithic age started with crude stone tools and ended with sophisticated big-game hunting. It supported a high but precarious standard of living. The Mesolithic age was harsher because environmental changes led to declining supplies of food; the dwindling resources gave impetus to innovation, so new technologies developed such as the manufacture of boats and the use of bows and arrows. The Neolithic age brought agriculture, the domestication of animals, and ultimately urbanization. These broad brushstrokes depicting preliterate history are painted from a palate of rich and diverse archeological discoveries. With a little imagination we can see the increasing role of reason as our ancestors emerged, explored, and adapted in a rapidly changing environment.

Stone tools are the signposts of hominid evolution. They have been found over 100 kilometers from their rock sources, which gives some indication of the nomadic life style of our forebears. Stone cutting edges were used to slice meat off bones and to make other tools from bone, ivory, antler, and wood. The combination of tools, manual dexterity, language, and reason created something new—a culture. The most pressing need for early cultures was food; the start of cooking, some 400,000 to 300,000 years ago, was thus a major step because it led to a vast increase in what could be digested.

We know from bones found in caves and burial sites that big game such as elephant and rhinoceros was caught. Torralba, Spain, has a hominid site with an assortment of animals' bones, some mixed with pieces of charcoal; the bones came from some thirty elephants, twenty-five horses, twenty-five deer, ten oxen, and several rhinoceros. Precise and consistent cooperation and communication would have been essential for the formidable task of hunting big game. Plans must have been formed to drive the prey into a bog or pit or over a precipice. Once dead, the animal was cut into pieces small enough to handle. By 14,000 years ago our ancestors were building fish traps for salmon, and by 10,000 years ago they were using bows and arrows.

REASON INCREASED THE precision and complexity of the stone tools that made everything possible, and reason was continuously and directly involved in the development of cooking, big-game hunting, and fishing. Reason grappled with the problem of finding shelter, for the climate over the last million years has undergone major changes. There were times when the temperatures in France and Germany were so high that tropical animals were swarming through dense forests. At other times, the extension of glaciers was accompanied by temperatures so low that hominids could survive only in caves. In general, the environment was colder than now, and primitive huts were built 500,000 to 300,000 years ago from clay, wooden branches, large animal bones, and animal skins. For long periods extensive areas of ocean froze so that sea level fell by some forty meters and land bridges appeared between the continents. These geographical changes played a fortuitous but vital role in allowing the dispersion of hominids over the globe; for example, land bridges during the ice ages allowed *Homo sapiens* to cross from Asia to Alaska.

Although food and shelter were the bare necessities of life for our ancestors, art also has a very long history. Ivory carvings of human and animal figures date back some 40,000 years. Soft stone and ivory female figurines were carved 24,000 years ago, and pictures of animals were painted on rock 19,000 years ago. The Chauvet cave in southern France contains some three hundred paintings of an assortment of mammoth, bison, horse, deer, woolly rhinoceros, bear, panther, hyena, and owl. Painted pottery appeared in both Europe and Asia 12,000 to 7,000 years ago. By 6,000 years ago, artists were painting pictures on cloth, showing people rowing boats on the Nile.

What was the purpose of all this ancient art? Rock painters often chose remote areas where access presented physical dangers, such as crossing subterranean streams or negotiating constricted passages, so the work does not seem to have been designed for the entire community. S. G. F. Brandon concludes

> that the motive behind the Paleolithic cave paintings was undoubtedly magical, and that it sought to capture or kill the various animals depicted by means of the magical efficacy of their painted substitutes. Now, if such was the motive of Paleolithic man in making these pictures, we may, reasonably, go on to speculate about the mental outlook which informed, perhaps unconsciously, the undertaking. This outlook may certainly be characterized as proleptic. In setting out to depict these animals, the ancient artist was surely looking ahead to the occasion of some future hunt. . . . (14).

Brandon cites the Dancing Sorcerer of the cave in the Trois Frères to support his views. This human figure, clad in animal skins and adorned with the antlers of a stag, is painted high on the wall and seems to dance over and dominate the animals below. The pictures were, perhaps, a kind of storage medium from which supernatural forces could be invoked whenever required. If there is any truth to these speculations, art is a product of social responses to a need for communication beyond the level of language, and often beyond the level of the natural world. Preliterate art seems to resonate with the supernatural—with magic and religion.

Early cultures consisted of small bands of nomadic hominids who hunted and gathered for some two million years. Over this time competition between different groups must have heightened the pressure for the evolution of more powerful brains. The archeological evidence is a long record of sudden death and dismemberment; indeed, *Homo sapiens* may have exterminated Neanderthals in warfare. One ingredient for success in combat would have been intellectual superiority, because cunning—the military equivalent of

Machiavellian intelligence—led to victory. The apt term for those who plot military deception is still "the intelligence service."

TOWARD THE END of preliterate history the human way of life underwent drastic changes over an action-packed period of 10,000 years. The transformation of small simple communities to large complex societies occurred at about the same time in Egypt, Mesopotamia, India, China, Central America, and South America. How did these changes take place? Reason can be seen in the background, actively facilitating the process. The first step, in abandoning the nomadic life, was settlement in fertile agricultural areas along rivers, lakes, and coastal strips. Some 12,000 to 6,000 years ago, wandering bands began to plant crops. Villages consisted of a few hundred people who farmed together and stored their produce communally. Goats and sheep were domesticated some 9,000 years ago, at a time when it was important to be able to graze on rugged terrain. Cattle and pigs were tamed 8,000 years ago, and horses 6,000 years ago. Dogs have been man's best friend and hunting companion for at least 12,000 years; recent evidence suggests that they may even have been with us much longer, perhaps over 100,000 years. The origin of our uniquely long and enduring relationship with dogs is beautifully portrayed in a Native American folktale:

> The earth trembled and a great rift appeared, separating the first man and woman from the animal kingdom. As the chasm grew deeper and wider, all the other creatures, afraid for their lives, returned to the forest—except for the dog, who after much consideration leapt the perilous rift to stay with the humans on the other side. His love for humanity was greater than his bond to other creatures, he explained, and he willingly forfeited his place in paradise to prove it *(15)*.

In early settlements most people were related to each other, and everyone needed a wide range of skills for survival. Contact with neighboring villages was limited to trading by barter, but from time to time the gene pool would be enriched by interbreeding. Disputes occasionally led to fighting between villages. Gradually, alliances formed to create larger and more powerful communities. In this way the small settlements coalesced, and some 5,500 years ago scattered dwellings at a few favorable sites became organized in rows to form streets. The size and complexity of communities grew as city-states evolved, with specialization and diversification of responsibilities among the population. Again, reason made these changes possible.

The knowledge accumulating from years of experience was memorized, codified, and passed from one generation to the next by the customs and ceremonies of oral tradition. Reason nurtured the growth of early cultures. It was indispensable for the discovery and operation of empowering technologies such as farming, irrigation, construction, transportation, the smelting of metals, and warfare. Once city-states were established, a strong centralized government became necessary if the community was going to prevail in the competition that would emerge with other city-states; reason played its part in the operation of government.

WRITING APPEARED SOME 5,000 years ago and this step provided an arbitrary but convenient criterion for drawing a line between epochs.

The invention of writing was driven by social forces—the need to improve communication between people separated by space and time. But writing proved to have powers beyond the social convenience of communication. Writing allowed information to be codified in permanent records. The new literate age built institutions dedicated to the acquisition, storage, and transfer of knowledge—libraries, schools, and ultimately universities. As a medium for communicating thought, writing promoted innovation; it played a central part in virtually all intellectual activity. Written letters became integrated with written numbers in new systems of mathematics. Reason was entrenched in these changes, which propelled human beings into an increasingly dominant position on the planet by help-ing them to accumulate knowledge.

Although it has helped us to survive, reason has not pointed us in any particular direction. Other influences have come into play, often unexpectedly. The increased size of the city-state brought new prob-lems: infections became a hazard, because city-states attracted vec-tors of disease such as rats, flies, and mosquitos. Previous small, wandering communities were too mobile for infections to become endemic; nomads left their excrement behind and moved on, whereas large static populations soon found that sanitation was a problem. With urbanization, traders moved from one city-state to another, spreading bacteria, viruses, and parasites, so populations developed high gene frequencies for traits (such as hemoglobin S and deficiency of glucose-6-phosphate dehydrogenase) that con-ferred resistance to endemic diseases. An efficient immune system became essential for survival, but sometimes infective disease could still ordain the rise and fall of empires. The smallpox virus proved a stalwart ally for Europeans setting out to conquer America. In such ways our past has unfolded with unexpected twists and turns. In the words of Alfred North Whitehead:

New epochs emerge with comparative suddenness, if we have re-gard to the scores of thousands of years throughout which the complete history extends. Secluded races suddenly take their places in the main stream of events: technological discoveries transform the mechanism of human life; a primitive art flowers into full satisfaction of some aesthetic craving: great religions in their crusading youth spread through the nations the peace of Heaven and the sword of the Lord (16).

EXISTING CULTURES
The term "culture" embraces all aspects of the way a community lives, including its language, morality, religion, art, government, law, industry, economy, recreation, food, and health. "Culture" also im-

plies a way of life that has been learned from the social and physical environment, as opposed to one that has been inherited. Cultures have exploited and promoted reason, but they have also repressed it, by burning rational books and persecuting rational people because rational views have often conflicted with powerful nonrational cultural beliefs.

Over the last half century, analogies have been drawn between cultures and living organisms. Cultures and organisms both have a complex structure with specialized parts contributing to the good of the whole, and they both have to adjust to a changing environment. Just as organisms have evolved through mutations that lead to new genes being passed from parents to their offspring by cell division, cultures undergo change through the introduction of new information (concepts, observations, or values) that is transferred to later generations by narrative and custom. As with most analogies, the likening of cultures to living organisms may lead to some useful insights, but it breaks down if it is pressed too far. The differences between the disciplines of biology and anthropology reflect the differences between organisms and cultures.

Anthropologists have always been intrigued by the diversity of human culture. For instance, in most parts of the world, marriage is a relationship based upon special sexual rights and the legitimization of children. But in Lesu, New Guinea, "The marriage relationship is symbolized by eating together. When a couple publicly shares a meal, henceforth they can eat only with one another. Even though husband and wife may have sexual relations with other individuals, they may not eat with them" (17).

A striking example of the difference between cultures surfaced during the Second World War. The Allies and the Japanese fought each other fiercely, each with total commitment to their cause. Allied soldiers who were taken prisoner maintained their allegiance to their countries and, in keeping with their tradition, cooperated minimally with their captors. Japanese prisoners of war, in contrast but also in keeping with their tradition, transferred their allegiance to their captors. In her classic account of Japanese culture, *The Chrysanthemum and the Sword*, Ruth Benedict describes how Japanese prisoners asked to be killed, "but if your customs do not permit this, I will be a model prisoner" *(18)*. They could certainly be model prisoners: old army hands and longtime extreme nationalists revealed the disposition of Japanese forces to the Allies and even flew with Allied bomber pilots to guide them to military targets.

Why is there such diversity among cultures? Why does our species behave in such different ways? One obvious explanation is the need to adjust to a wide range of local environments. Snow-land communities devise warmer clothes, improved methods for constructing shelters, and mythologies that express respect for local predators such as bears and wolves. All of this would be decidedly inappropriate in the tropics. Thus there is a definite rational component to the diversity of cultural practices. But cultures also make capricious, irrational choices. There is no rational explanation for why Christians take their hats off to pray while Jews put theirs on, or why Hindus won't eat beef while Muslims won't eat pork.

In addition to these rational and irrational differences between cultures, there are rational similarities. The same behavioral threads are woven together to produce a rugged social fabric. Donald Brown has termed this survival kit of cultural similarities "human universals": *Homo sapiens* always lives in groups; the groups have a language and they recount traditional stories; there is a ranking of prestige within the group; social manipulation takes place, and there is a triangular awareness—individuals interpret how others interact and they infer the relevance of such behavior to themselves; plans are devised for future activity; ownership, implicit in territoriality, is extended to objects; government is installed; ritual and religion are established; codes of ethics are created and laws are enacted; intention and responsibility are recognized; standards of courtesy are formalized; alliances are formed between individuals; marriage is practiced;

copulation is not undertaken in public; the core of the family is the biological mother and her children; members of a family support each other; children play with each other and learn their cultural customs at an early age *(19)*.

Every now and then an aberration appears in the rational infrastructure of cultures. For instance, all existing cultures discourage incest. This taboo has natural biological value because of the genetic disadvantages of inbreeding. But there have been times when incest was practiced at the most elite levels of society, without recognition of the unwanted consequences *(19)*. Shortly after the conquests of Alexander the Great, a Greek king of Egypt divorced his wife to marry his sister; this seemed to set a fashion, for seven of the next eleven Greek kings of Egypt married their sisters. When the Romans conquered Egypt they conducted a census revealing that at least 15 percent of reported marriages were incestuous. So from time to time there have been odd deviations from the rational underpinning of cultural universals.

All cultures are built upon inborn patterns of behavior that we share. For example, one kind of genetically determined building block is the response to certain smells. Attraction to the smell of food is obviously useful for survival, and aversion to the smell of excrement reduces the risk of contracting infective disease from feces and urine. More complex cultural universals are recapitulations of facets of behavior in the social life of monkeys and apes, but in *Homo sapiens* we find increased versatility, with language helping to expand the range of options.

We have discussed reciprocal altruism in animal communities; it is equally embedded in the operation of human societies. In the First World War, the carnage seemed so senseless that some sectors of the front established a practice of firing their artillery barrages at regular and therefore predictable times. Each side recognized when to expect their opponents' shells so they knew when to go into their dugouts. This mutually beneficial arrangement reduced loss of life and complied with the need to demonstrate the consumption of large-caliber ammunition.

SOCIAL MOTIVATION

People identify themselves as belonging to groups beyond their families—nation, race, religion, class, profession, gang—and social motivation is driven by shared beliefs set in the values of the group. Such motivation can, under suitable circumstances, dominate all other considerations. Wars reveal the extremes of human behavior that cultures are capable of imposing. Throughout history men and women have laid down their lives in willing support of their religious or political ideologies. Religious and political leaders have demanded the remorseless sacrifice of young men in an attempt to demoralize the enemy. Cultural forces, channeled through the emotions, give the incentive to fight while reason enables the machinery of war to become increasingly efficient. Instead of hundreds of people dying in a war, the toll has reached millions as rational technology creates more powerful machine guns, artillery, and nuclear bombs. But reason remains morally neutral, for it can equally be applied to the task of sustaining life through medical science, improved food production, and the distribution of cheap energy. Reason can be used by whoever grasps it, for whatever purpose—it does not discriminate.

AS AN ILLUSTRATION of how irrational cultural attitudes drive massive armies against each other we can pick one of the numerous battles fought in the First World War. The British planned to break the deadlock on the Western front with an offensive mounted from a sector on the River Somme. On the first day, July 1, 1916, 60,000 British soldiers were missing in action. By the middle of November, there were 419,654 British casualties. The magnitude of these figures, for a single battle fought over four months, can be placed in perspective by comparing them with the 50,000 American dead and the 160,000 wounded over the entire ten years of the Vietnam War.

Shortly after the start of the First World War, Lord Kitchener called for a single increment of 100,000 volunteers to add to the British regular army. Within a few months he had 600,000.

How do cultures impose their irrational wills so easily? Nothing significant was gained in return for the appalling British losses on the first day on the Somme, but the policy to continue the attack did not falter. The only positive comment that can be made about this epic disaster is that the world has seldom witnessed such an enormous collective display of bravery. John Keegan describes the volunteer army that fought the battle, predominantly young men who enlisted because of their sense of belonging to Britain and

> to work-places, to factories, to unions, to churches, chapels, charitable organizations, benefit clubs, Boy Scouts, Boy's Brigades, Sunday Schools, cricket, football, rugby, skittle clubs, old boys' societies, city offices, municipal departments, craft guilds—to any one of those hundreds of bodies from which the Edwardian Briton drew his security and sense of identity.... Many surrendered well-paid steady employment to join up, coming forward in such numbers that they overwhelmed the capacity of the army to clothe, arm and train them.*

Units gave themselves regimental subtitles that reflected the close ties of small communities: "North-east Railway, 1st Football, Church Lads, 1st Public Works, Empire, Arts and Crafts, Forest of Dean Pioneers, Bankers, British Empire League, Miners." Most of the men fought because they saw it as their duty. Most had no thoughts about the politics or purposes of the war, and they had no concerns about the competence or intention of their leaders. The courage of the British Expeditionary Force was unsurpassed. Keegan reports an

---

* John Keegan, *The Face of Battle* (New York: Viking Press, 1976). Keegan, a British military historian, gives a penetrating description of armed conflict, taking as pivotal examples Agincourt, Waterloo, and the Somme.

eyewitness account just before an attack, given by the commanding officer of the 9th Royal Inniskilling Fusiliers, who

> stood on the parapet between the two centre exits to wish them luck. . . . They go without delay, no fuss, no shouting, no running, everything solid and thorough—just like the men themselves (these were farming people from County Tyrone). Here and there a boy would wave his hand to me as I shouted good luck . . . through my megaphone. And all had a cheery face. Most were carrying loads. Fancy advancing through heavy fire with a roll of barbed wire on your shoulder!

The losses of junior officers were particularly shocking; leaders in the attacks were the first to fall. Donald Hankey, who himself died on the Somme, wrote an essay for the *Spectator* on "The Beloved Captain." "We knew we should lose him . . . but how was the company to get on without him? To see him was to forget our personal anxieties and only think of . . . the regiment and honour."*

Very occasionally the stress became intolerable, and soldiers did ask why they should die in such senseless circumstances. The prevailing culture dictated the outcome for those who wavered. Keegan describes what happened when a group in the Ulster Division decided to leave an attack that was encountering heavy fire. "They were damned if they were going to stay. . . . a young sprinting subaltern heads them off. They push by him. He draws his revolver. . . . They take no notice. He fires. Down drops a British soldier at his feet. The effect is instantaneous. They turn back. . . ."†

Japanese culture demanded similar self-sacrifice in the Second World War. The Imperial Japanese Army codified its expectations in its publication *Ethics in Battle.*

---

* Hankey is quoted by John Keegan as an example of a cadre of religious officers, who saw the virtues required by the perfect officer as resembling those of Christ.
† Cultures know very well how to combine the carrot of honor with the stick of disgrace—accolades for heroism and summary execution for cowardice.

A sublime sense of self-sacrifice must guide you throughout life and death. Do not think of death as you use up every ounce of your strength to fulfil your duties. Make it your joy to use every ounce of your physical and spiritual strength in what you do. Do not fear to die for the cause of everlasting justice. Do not stay alive in dishonor. Do not die in such a way as to leave a bad name behind you (20).

General Sakurai expressed the military culture explicitly in a 1944 order of the day. "If your hands are broken, fight with your feet. If your hands and feet are broken, fight with your teeth. If there is no breath left in your body, fight with your ghost" (20).

In 1944 the Japanese government recognized the deterioration of their military position in the Pacific. In response, they formed special attack groups, kamikaze ("the divine wind," named after a typhoon that destroyed a Mongol fleet attempting to invade Japan in the thirteenth century). Vice-Admiral Onishi commanded the special attack forces. On October 20, 1944, he addressed a group of twenty pilots:

Japan is in grave danger. The salvation of our country is now beyond the powers of the Minister of State, the General Staff, and lowly commanders such as myself. Thus, on behalf of your hundred million countrymen I ask of you this sacrifice and pray for your success. You are already gods without earthly desire. If you had such desires they would be whether or not your body-crashing efforts achieved success. You will not be able to know this because you will enter on a long sleep. We cannot let you know the results, either. I shall watch your efforts to the end and report your deeds to the Throne. You may all rest assured on this point (20).

Onishi had little concern about persuading the pilots to accept suicide missions; there was no shortage of volunteers.

By the end of the kamikaze attacks, 3,913 Japanese pilots had died. When they climbed into their cockpits they knew, for certain, that they would receive automatic posthumous promotion; they also knew, for certain, that they would not return. Many were high school students, some of them no more than seventeen years of age. Many more were preparatory trainees, or apprentice pilot officers. Before they went, they sent home *haiku* to their mothers. Perhaps "The Hymn of the Dead" expressed their feelings best:

If I go away to sea
I shall return a corpse awash.
If duty calls me to the mountain,
a verdant sward will be my pall;
For the sake of the Emperor I will not die
peacefully at home.

The kamikaze pilots displayed bravery similar to the British soldiers who died on the Somme. Valor is heroic, but there is a darker side to cultures that inspire their members to feats of extraordinary bravery. Japanese nationalism generated men who were fully prepared to die, but it also led to horrific eagerness to kill. An eyewitness account depicts how a Japanese surgeon, Major Komai, executed a captured Allied intelligence officer in New Guinea:

The Major draws his favourite sword. It is the famous *masamune* sword which he has shown us at the observation station. It glitters in the light and sends a cold shiver down my spine. He taps the prisoner's neck lightly with the back of the blade, then raises it above his head with both arms and brings it down with a powerful sweep. I had been standing with muscles tensed but in that moment I close my eyes.

A hissing sound—it must be the sound of spurting blood, spurting from the arteries: the body falls forward. It is amazing—he has killed him with one stroke.

The onlookers crowd forward. The head, detached from the trunk, rolls forward in front of it. The dark blood gushes out. It is all over. The head is dead white, like a doll. . . . A corporal laughs: "Well—he will be entering *Nirvana* now." A seaman of the medical unit takes the surgeon's sword and, intent on paying off old scores, turns the headless body over on its back and cuts the abdomen open with one clean stroke *(21)*.

The juxtaposing of bravery and brutality, apparent when we compare the deeds of kamikaze flights with this execution scene, illustrates the strength of the imperatives that can be generated by a culture. What is the biological role of the cultural drive to sustain such enthusiasm to die and to kill? United and determined societies are likely to dominate those made up of dissenting groups with weak policies and little commitment. But the absence of criticism within the culture carries dangers. If a highly disciplined nation is losing a war, failure is disastrous. The culture becomes totally discredited, so its wholesale destruction and replacement are inevitable. A more loosely structured culture may not be as efficient as a fighting machine, but it has the advantage of greater flexibility; by being open to negotiation, it is more likely to survive defeat.

WHY DID THE most powerful cultures of the twentieth century propel themselves into such disastrous contests? Both world wars were clashes of imperial might, or more specifically, conflicts over the control of resources—territory, raw materials, and markets. The issues were of the same type as those determining the behavior of groups of monkeys and apes. The soldiers were victims of the power wielded by unyielding cultures that dictated to each man where his allegiance lay, what his obligations were, and how he should fulfill them. If there were any doubts, leaders had no compunction in extinguishing them.

The behavior of nations at war helps us to understand the forces available to a culture when it becomes focused on a single issue—victory at any cost. In times of war, reason is recruited to plan better tactics, build faster airplanes, and design more powerful bombs. Reason is suppressed where it might jeopardize the war effort by questioning self-sacrifice. Soldiers are trained to "do and die" rather than "reason why"—they are ordered to behave like driver ants on a raid, and they obey.

THE NEUROLOGY OF SOCIAL BEHAVIOR

The evolution of cultures took place while several regions of the brain were expanding. Which parts of the brain are primarily responsible for social behavior? The answer comes from observations on patients who have lost their social skills through localized damage within the brain. The case of a railroad construction foreman gave us the salient facts in 1848. Phineas Gage, aged twenty-five years, was laying track in Vermont when an accidental explosion blew an iron bar through the front of his head. Antonio Damasio has recently conducted a detailed analysis of what happened (22). Before the explosion Gage was reported by the railroad company as being "the most efficient and capable" man in their employ. The circumstances following the accident are graphically recorded: the explosion threw Gage onto his back and he had a "few convulsive movements of the extremities." He spoke a few words and was carried to an ox cart; he was then taken, sitting, to the nearest town, where he received prompt medical attention. He made an excellent recovery, but he displayed a remarkable change in behavior. His physician, Dr. Harlow, described him as "fitful, irreverent, indulging at times in the grossest profanity which was not previously his custom, manifesting but little deference to his fellows, impatient of restraint or advice when it conflicts with his desires." Damasio comments that "the foul language was so debased that women were advised not to stay long in his presence, lest their sensibilities be offended. The strongest admonitions from Harlow himself failed to return our survivor to good behavior." In essence, the iron bar had damaged Gage's faculty for social skills; he was no longer able to recognize or inhibit behavior that was unacceptable to others. Gage died in 1861 and his skull has been preserved. Damasio and his colleagues have performed reconstructions that reveal the trajectory of the iron bar. It passed up the middle of the front of the brain, inflicting serious injury to the inner aspect of both cerebral hemispheres and sparing the outer regions. Damasio identifies the critically affected portion of the brain as the ventromedial frontal area (the lower part of the frontal lobe, near the midline). He concludes: "The unintentional message in Gage's case was that observing social convention, behaving ethically, and making decisions advantageous to one's survival and progress require knowledge of rules and strategies and the integrity of specific brain systems." Subsequently, many patients with similar problems have been studied, all with frontal brain lesions leading to difficulty in comprehending their own social role and in interpreting the behavior of others.

With this background, experimental lesions have been placed in the front of the brain in monkeys. The procedure results in loss of social position within the monkey troop. The monkeys display less social activity, such as grooming, vocalization, and for the females, maternal behavior; they seem unable to recognize and practice the social conventions required by the troop.

So neurological evidence points to the frontal region of the brain as having special responsibility for social behavior. Does this help us in our attempt to unravel the origin of reason? Compared with the brain of an ape, the human brain has many expanded areas, but by far the greatest enlargement is in the frontal region. This finding, together with the reports on Phineas Gage and similar patients, supports Humphrey's suggestion that the evolution of intelligence was linked to the evolution of social skills. Successful social interaction must have given early hominids an advantage in survival so that enhancement of the "social brain" became a trump card in natural selection. The high level of frontal development in *Homo sapiens* has contributed to the spectacular success and proliferation of the species.

Stanley Garn summarized the implications of these conclusions.

He emphasizes the importance of the "social brain" (responsible for human cooperation), as opposed to the "tool using brain" (responsible for manual dexterity):

When we look upon the social adaptations that man has made, adaptations to the improbable rules of his own making, when we observe (in even simplest societies) the highly complicated game of human relations, we wonder about the alterations in brain structure and brain physiology necessary to make such behaviors possible. The necessity to adjust to complex relationships with an ever-changing set of age-specific rules, the needs for imagining, worrying, dreaming and even speculating seem more closely related to our direction of brain evolution than simply using tools *(23)*.

Because of the linkage between social skills and intelligence, destruction of the frontal lobes should lead to disturbances that extend beyond loss of social grace. Tests for "decision making" have been developed and, sure enough, when normal people are compared with patients who had damage to the frontal lobes, the normal subjects are better at making advantageous decisions *(24)*. Furthermore, the large size of the frontal lobes allows specialization of function within their different parts. While lesions of the ventromedial frontal region lead to diminished social skills (the disinhibition syndrome), lesions of the dorsolateral portion, situated at the top of the frontal lobes toward the sides, result in decreased mental agility and difficulty in assembling thoughts (the dysexecutive syndrome) *(25)*. A third constellation of symptoms is called the apathetic syndrome—patients lost interest in themselves and their surroundings, without being depressed. Here, damage is localized to the inner aspect of the frontal lobes, in the cleft between the two hemispheres. The pathways for nerve fibers passing to and from the frontal areas are complex chains of nerve cells that form independent parallel loops connecting the frontal cortex to various deep structures in the brain. Five such circuits have been identified, and there are probably more *(26)*.

FROM THESE OBSERVATIONS we can see how the evolution of the frontal lobes was an essential step in the development of human reason. The most obvious frontal responsibilities—social skills and the ability to assemble and adjust a series of thoughts to reach a goal—provide the scaffolding for reason. It would, however, be a mistake to envision reason as "locked up" in the frontal lobes. Pathways to and from other regions of the brain must also contribute because reason operates in tandem with other cerebral modules, such as language and memory.

5

ETHICS

The distinction between right and wrong is made by people
on the basis of how they would like their society to function.
It arises from interpersonal negotiation in a particular
environment, and derives its sense of obligation and
guilt from the internalization of these processes.
Moral reasoning is done by us.

—FRANS DE WAAL*

SHOULD WE HAVE the death penalty? Should we permit abortion? Should society allow assisted suicides? People have diametrically opposed views on these issues. Who is right? We have no easy answers, but there are questions of this type that draw general agreement: thou shalt not kill; thou shalt not steal; thou shalt not bear false witness; thou shalt not commit adultery (or more generally, violate sexual customs). In spite of the wide diversity of existing cultures, they speak as one when it comes to these fundamental standards of human behavior.

---

* Frans de Waal's *Good Natured* (3) gives a beautiful account of the gentle and generous behavior of animals—aspects that we have tended to ignore or deny. This book is particularly important because it offers a persuasive view of how our own ethical values evolved, and in so doing it tells us much about the nature of our morality.

How are these "ethical universals" established? The word "ethics" comes from the Greek *ethos,* and "morality" from the Latin *mores;* both terms are used to discuss how we should behave to each other, and more precisely, how we can reconcile the eternal problem of finding ways to cooperate in a community, where individuals will never be able to get everything they want. In the words of Geoffrey Warnock:

> Resources are limited; knowledge, skills, information, and intelligence are limited; people are often not rational, either in the management of their own affairs or in the adjustment of their own affairs in relation to others. Then, finally, they are vulnerable to others, and dependent on others, and yet inevitably often in competition with others; and, human sympathies being limited, they may often neither get nor give help that is needed, may not manage to co-operate for common ends, and may be constantly liable to frustration or positive injury from directly hostile interference by other persons *(1).*

Because everyone cannot have everything they want, we need ethical principles to find a middle ground between, on the one hand, the callous pursuit of self-interest and, on the other hand, repression by enforced social conformity.

Does reason influence the way cultures decide what is moral? This is a central question in our effort to define the role of reason. Modern cultures have established codes of behavior that can usually be traced to religious origins, in particular to revelation by a prophet. The world's major religions arose thousands of years ago in societies that were culturally, racially, and geographically widely separated, yet they have all produced very similar guidelines on how people should behave toward each other. Is this coincidence, or is reason the common denominator that unifies the moral principles taught by the different religions?

The evidence suggests that morality is a coherent system of rational guidelines on how communities have to operate if they are to survive in an environment with limited resources. In short, reason underpins what we call good and evil, right and wrong, just and unjust. Is ethical reasoning conducted in the same way as the study of facts about the nature of the world? P. H. Nowell-Smith considers that the commonest error in moral philosophy "is that of transferring to discussions of moral discourse the logical concepts that have been successfully used to elucidate the discourse of mathematics or science" *(2)*. If Nowell-Smith is right, we have an odd inconsistency in the application of reason; but perhaps he is wrong. A rational backbone to morality (in the same sense that science has a rational backbone) becomes plausible if we recognize that ethics are not just a supercode of idealized answers to theoretical questions of what is good or bad. They have a very practical biological purpose, namely *to reconcile the needs of society with those of the individual*—and reconciliation is a necessary condition for the survival of our species. If this is the purpose of morality, then different approaches to ethical issues can be judged by the extent to which they succeed or fail, and ethics can be subjected to the normal methods of rational inquiry that are used for exploring other natural phenomena. In other words, reason *can* be used in moral discourses, in the same way that it is used in mathematics or science.

BIOLOGICAL ORIGINS OF ETHICS

If *Homo sapiens* had been a solitary creature there would have been little need for language to evolve, and the biological value of ethics would scarcely have extended beyond conventions to facilitate courtship, mating, and rearing. Ethics would be diminished in stature to etiquette. But it is clear that *Homo sapiens* is gregarious, forming large groups that operate complex, coordinated societies. For this gregariousness to work, each group must have a social structure with rules to determine the placement of individuals, and how they should interact. Hitherto, biological opinions have stressed competition as a vital force in determining survival of both the species and the individual. In monkeys, well-described hierarchies rank the stronger members over the weaker, and this pattern of social dominance seems to satisfy evolutionary needs for survival of the fittest. The "strongest" generally means the most powerful physically, determined when necessary by contests, but cunning can also be used for social purposes. Hence the idea of Machiavellian intelligence gained favor to explain how physically weak but mentally strong individuals can assert themselves.

Yet community life in animals imposes restraints; for instance, fights to determine leadership end when the loser submits and accepts the result. Why doesn't the winner press his advantage and destroy his competitor once and for all? The answer is to be found in Frans de Waal's recent challenge to the traditional view of "nature red in tooth and claw" (3). He disagrees with the prevailing opinion that attributes selfish and cruel motives to animals, starting with a forthright statement denouncing the zealots who castigate any attempt to dwell on kind or generous interpretations of animal behavior:

> As a student of chimpanzee behavior, I myself have encountered resistance to the label "reconciliation" for friendly reunions between former adversaries. Actually I should not have used the word "friendly" either, "affiliative" being the accepted euphemism. More than once I was asked whether the term "reconciliation" was not overly anthropomorphic. Whereas terms related to aggression, violence and competition never posed the slightest problem, I was supposed to switch to dehumanized language as soon as the affectionate aftermath of a fight was the issue. A reconciliation sealed with a kiss became a "postconflict interaction involving mouth-to-mouth contact"* (3).

---

* Animals who live in communities frequently help each other. Though they fight with each other to compete within the group, they will unite to fight more aggressively against another group. The similarities to human behavior are compelling.

De Waal emphasizes the importance of friendship, citing an example from his first view of a chimpanzee's birth. A crowd of chimpanzees are watching with him.

> After approximately ten minutes, Mai tensed, squatted more deeply, and passed a baby, catching it in both hands. The crowd stirred, and Atlanta, Mai's best friend, emerged with a scream, looking around and embracing a couple of chimpanzees next to her, one of whom uttered a shrill bark. Mai then went to a corner to clean the baby and consumed the afterbirth with gusto. The next day Atlanta defended Mai fiercely in a fight, and during the following weeks she frequently groomed Mai, staring at and gently touching Mai's healthy new son *(3)*.

Similar acts of friendship are frequent in many different types of animal community, although they have been ignored until recently. Unselfish behavior extends beyond monkeys and apes: dolphins support injured members of the group by swimming under them, and may even beach themselves in a vain attempt to help. Predators bring food back to those who were not in at the kill. Is there a biological advantage for friendship? Surely there is, and it lies in the creation of social cohesion; a coordinated group who help each other will, in general, operate more effectively in the struggle for survival than isolated individuals driven solely by immediate self-interest.

If human morality grew from the social behavior of animals, we may be able to gain some better understanding of ethics by looking at the relevant traits in chimpanzees. De Waal has suggested that there are four *(3)*. First, there is the trait of "sympathy" that comes from an ability to identify with another individual, to change places mentally. Second, there is the trait of "internalization," the capacity to generalize from specific experiences so that a pattern of rules can be recognized, and punishment can be expected for breaking them. Third, there is the trait of "reciprocity," the generalized expectation of returned social treatment. Fourth, there is the trait of "getting along" with each other and avoiding conflict. A synthesis of these four features of animal behavior offers a blueprint for morality, and it also illustrates the essential part played by reason, for each of the four traits is a rational response to the goal of balancing competition with cooperation.

WITH THIS BACKGROUND we can speculate on the behavior of early hominids. We know they formed wandering bands, and they had language and skills in hunting and gathering. These small groups must have built an informal social structure for making joint decisions, and a code of behavior would have been necessary, even if this was little more than the sort of understanding that arises in an extended family—the idea that some conduct is acceptable and some is not. As the bands grew in size they became more powerful—provided they could function as a group rather than a collection of individuals. The only way to keep cohesion was through a system of regulation, or governance, with laws based upon a morality. What people thought, as a community, became important and "public opinion" had to be respected. To ensure that ethical standards would be accepted by everyone, it was necessary to develop guidelines that seemed fair in the light of people's circumstances and expectations. The code of behavior gained further strength if its origin was assigned to divine authority.

For a community to prosper, reason demanded that its ethical code had to control, at the very least, physical violence, the ownership of property, sexual practices, and communication. Of course, reason was not applied in any formal way to the task of creating ethical standards, but at every twist and turn in the building of an operational morality, our ancestors were using reason to reach decisions, correct mistakes, establish consistency, and ultimately deliver a system that worked. Individuals have to subordinate their competitive urges in order to cooperate, but how is this done? An ethical code has to define what *is* and *is not* socially acceptable—in other words, what is right and wrong. In an ever-changing society, the task of choosing between new options is, *par excellence,* a job for the faculty of reason. As we shall see, we use reason to decide that medical treatment should be given to sick people when they cannot afford it. We use reason to decide that pensions are necessary for people who are no longer productive in the community. We use reason to decide that every member of the community, however humble, has rights.

Having accepted that morality has its origins in the biological problem of reconciling competition and cooperation, we can now look at the historical record of how reason has shaped morality. We find that reason has been engaged repeatedly to analyze existing ethics in the hope of developing a better system, and here we have much firmer evidence than our speculation on the morality of pre-literate bands of wandering hunters and gatherers. Ancient civilizations in Greece, the Middle East, and Asia pioneered the rational analysis of ethics.

ANCIENT VIEWS ON ETHICS

From Greece to China, the early literate cultures were very much concerned with how people should live. Their approaches differed in emphasis and in detail, but they also had much in common. Egyptian and Babylonian texts, written 5,000 years ago, dealt with ethical issues in a very practical way. Members of the Egyptian ruling class were encouraged to be fair and generous; they were told that such conduct was both good and profitable, as explained by P. H. Nowell-Smith:

> The Egyptians had a word "Ma'at" which is translated in three different ways. It means (a) "being straight, level, or even," (b) "order, conformity, regularity," and (c) "truth, justice, righteousness." It has obvious affinities to our word "right." Now the earliest moral document we possess is a manual of instruction on good behavior to budding civil servants. In this the aspiring official is enjoined to follow the rules of Ma'at, but only because, by so doing, he will get

on in the world. . . . We are to practice virtue because, in the short
or long run, it pays *(2)*.

The Babylonians believed in reciprocity—an eye for an eye, and a
tooth for a tooth, though a master's eye was worth more than a
slave's. Ancient Hebrew culture picked up these attitudes, but having
been enslaved themselves, Jews took a more egalitarian position. The
Ten Commandments arose from this background, as did the idea
that God would reward those who were good and punish those who
were bad. The mismatch—that good conduct did not always lead to
happiness or bad conduct to suffering—created a dilemma that was
only resolved later, by the idea of divine judgment and life after
death.

Early Indian writings on ethics were more philosophical than
practical; the Hindu Veda expressed what was true and what was
right as being the same. The universe entailed a moral order as part
of its nature. Jainism, an offshoot from Hinduism, emphasized non-
violence and the preservation of all forms of life. Buddhism, another
offshoot from Hinduism, swept across Asia with a cogent message
that compassion and friendship were more important than anything
else in life.

In China, Lao-tzu was born around 600 B.C., and although very little is known about him, his teaching of the virtues of simplicity and
sincerity culminated in a way of life called the *Tao*. This was to become codified in the *Tao Te Ching*, or *The Way and Its Power*. The
most influential moral philosopher was Confucius, who followed
Lao-tzu. Confucius was among the first to argue the golden rule of
ethical reciprocity: "Do not do to others what you would not wish
done to yourself." Confucius encouraged integrity in a practical way,
achieved through deeds, so that the community would respect your
name after you die.

The Greeks of the fifth and fourth centuries B.C. were the trail-
blazers in early attempts to apply reason formally to ethical questions. The Sophists taught debating skills professionally, and they

made their own contributions to philosophy. Protagoras was the best known; he was a friend of Pericles, who invited him to prepare a legal code. Most of the works of Protagoras have been lost, but one famous maxim has been passed down: "Man is the measure of all things." Protagoras held that there is nothing absolute about ethical concepts—they are all relative. What is "good" and what is "bad" depend upon social conventions and therefore vary from place to place and time to time.

Socrates also applied reason to ethics, but he disagreed with the Sophists; he believed it was possible to know what was good in an absolute sense. He considered that people who know what is good must necessarily behave in a virtuous way, and that what is good for the individual is good for the community. The most distinguished disciple of Socrates, Plato, agreed with the idea of absolute ethical standards and argued that there is some general and pure form of goodness, though it is not easy to define. In the *Republic* the "good life" is linked to rationality. Socrates and Plato believed in immortality, and they were quite comfortable with the notion that people should do good in order to benefit themselves in a life after death. Aristotle did not accept this "essence of goodness"; he developed another rational view of ethics—things are good if they fit their purpose well—but he did not give a clear description of what moral purpose might be.

Before, during, and after this time, various Greek philosophers argued for hedonism—that pleasure, in one form or another, is good. Aristippus, a contemporary of Socrates, favored immediate sensual pleasure as the ultimate good. His philosophy propelled him to the court of Dionysius in Syracuse; here Aristippus is reputed to have thrown himself, with enthusiasm, into the practice of what he preached.

In sharp contrast, Zeno of Citium founded what became known as the Stoic school of moral philosophy, named after its location at the *Stoa Poikile* (Painted Colonnade). The Stoics held that the only really important thing in life is virtue: "health, happiness, possessions are of no account. Since virtue resides in the will, everything re-

ally good or bad in a man's life depends upon himself. He may become poor, but what of it? He can still be virtuous. A tyrant may put
him in prison, but he can still persevere in living in harmony with
Nature. He may be sentenced to death, but he can die nobly" (4). The
Stoics stressed that reason must be used to decide what is good, and
since everyone has the capacity for reason, everyone is a custodian of
morality—in contemporary terms, morality had a democratic dimension.

The Epicureans were the last influential philosophers before the
massive changes that occurred when the Roman Empire embraced
Christianity. Epicurus (341–270 B.C.) took a disarmingly simple and
rational ethical stance: anything that increased pleasure or decreased
pain was, by definition, good—but he was not thinking of the immediate pleasures of the hedonists. Epicurus dismissed sensual pleasures quite firmly: "When we say that pleasure is the goal we do not
mean the pleasures of the dissipated and those which consist in the
process of enjoyment . . . but freedom from pain in the body and disturbance in the mind. For it is not drinking and continuous parties
nor sexual pleasures nor the enjoyment of fish and other delicacies
of a wealthy table which produce the pleasant life, but sober reasoning" (5). The ultimate good for Epicurus was the tranquility that
comes from eliminating unnecessary desires so that most needs
could be met for most people. In his own words: "We say that pleasure is the starting-point and the end of living blissfully. For we recognize pleasure as a good that is primary and innate. We begin every
act of choice and avoidance from pleasure, and it is to pleasure that
we return using our experience of pleasure as the criterion of every
good thing" (5). The rational arguments of the Epicureans anticipated many modern views. Sadly, only three letters and a few fragments of the works of Epicurus have survived, for he wrote some
three hundred lost volumes. Just as Aristippus had pursued a life in
keeping with his hedonist philosophy, Epicurus founded communities where his followers could escape to pursue a life of temperance
and simplicity.

MODERN VIEWS ON ETHICS

The history of ethics reveals how reason generated a variety of answers to ethical questions. Accumulating knowledge led to the reevaluation of old moral issues, and a need to look at new ones. From St. Augustine to John Rawls, theologians and philosophers continued to grapple with questions of what people should and should not do, and why. But solutions did not come easily; the strife and bloodshed over the present century shows how poorly we have learned to live together.

Fifty years ago nuclear physics seemed to pose ethical dilemmas, and now biology is at the center of the stage. Research on DNA and the resulting advent of cloning has obvious moral implications. Recently, biologists and economists have started to work on the practical results of different kinds of behavior, and they are discovering the benefits of long-term cooperation, as opposed to competition. The evidence is discussed at the end of the next chapter, because the finding that cooperation is unexpectedly advantageous has immediate relevance to commerce. Another new direction for ethics is the analysis of how our closest evolutionary relatives behave to each other, and, as we have seen, observations on the social behavior of monkeys and apes have enabled us to examine our own morality with fresh insight.

MICHAEL PHILIPS HAS recently integrated a philosophical and biological approach to the analysis of ethics (6). He starts with the premise that morality is a rational response to a biological problem. He points out that ethics enable a wide range of highly complex and often Machiavellian behavior patterns in *Homo sapiens* to be controlled for the benefit of the community—which in turn brings benefit to the individual. Ethics use reason to make aggressive, self-interested people into team players by engaging their social skills in the task of getting along with each other. Philips sets out to examine the old conflict between those who argue extreme skepticism (everything is relative) and those who argue extreme universalism (everything is absolute). He seeks a rational middle ground by embracing the biological view that ethics are created by people to keep their social structures from self-destructing. In his book *Between Universalism and Skepticism,* Philips uses the subtitle *Ethics as a Social Artifact.* He describes this theory of ethics in terms of a morality that is essentially instrumental. Morality is presented as a tool that can be evaluated by how well it does its job, and its job is defined as the task of reaching a social goal—the promotion of cooperation. Ethics as a social artifact contrasts with religious morality, in which ethics are judged by how well they conform to a divine supercode that tells us what is good. In the words of Philips:

> Every human society has certain fundamental needs. Among other things, it must care for and socialize its young, produce and distribute material goods, heal the injured and the sick, enforce its moral standards, and protect itself from attack. In every society, moreover, there is a division of labor to meet these needs. This division of labor leads to what I call domains. Roughly, a domain is defined in terms of a set of goods and evils and a set of roles. The goods and evils define the purpose of the domain, and the roles define the social structures that are supposed to realize those purposes* *(6).*

Everyone operates within a special subset of moral standards relating to their particular domain. For example, a physician's duties are summarized by the Hippocratic oath, and enforced by various professional colleges and councils. In addition, there is a more general "umbrella" of universal guidelines governing behavior, the

---

* Examples of domains are medicine, law, teaching, commerce, and construction work. Each of these domains is defined by qualifications, duties, rewards, and punishments.

core "commandments"—not to kill, steal, lie, or violate sexual customs.

This analysis is persuasive; it offers a comprehensive, coherent explanation of *why* we have a morality and *how* it operates. In particular, the ideas presented by Philips allow us to see the way the rational infrastructure of human ethics grew out of the social behavior of our closest living evolutionary relatives, monkeys and apes.

IMMORALITY

What happens when we break the commandments that are the moral foundations of our culture? To answer this question we can start by taking one of the commandments shared by almost every culture: thou shalt not bear false witness. A cursory glance at a range of cultures shows that lying is universally condemned, but that lying is universally practiced. While any deliberately untrue statement is a lie, there are clearly different kinds of lies: on the one hand there are those told for self-interest, and on the other, there are those told to help people. We have no qualms about lying if the truth is unnecessarily painful. When someone who is preparing to go out to dinner looks horrible and asks, "How do I look?," we are likely to lie, and we are unlikely to have the lie on our conscience— conscience being a repository of ethical values built up by the culture and inculcated into the young as a blueprint of morality which, if broken, results in remorse. Of course, the matter is often more serious than this, for lying can have profound repercussions. Sissela Bok has analyzed the implications of lying and cites the views of many authorities *(7)*. For example, St. Augustine was adamantly against any form of lying. In discussing whether we should lie to save others from unnecessary distress, he says no, lest "little by little and bit by bit this evil will grow and by gradual accessions will slowly increase until it becomes such a mass of wicked lies that it will be utterly impossible to find any means of resisting such a plague grown to huge proportions through small additions."* Martin Luther disagreed: "What harm would it do if a man told a good strong lie for the sake of the good and for the Christian church?"† By lying for the sake of others we may be breaking an ethical code, yet most of us do not feel shame because we interpret morality as flexible enough to be stretched this far. But the situation may not always be straightforward. When governments lie, who is to decide whether it is for the general good?

This kind of dilemma is common. By exceeding the speed limit on an empty road we are primarily risking our own lives, so few of us would feel guilty. Many drivers would hold that exceeding the speed limit in this way is neither immoral nor irrational—and others would disagree. Ascending the misdemeanor scale, what should a physician do when a lucid patient who is dying with intractable pain asks for a lethal injection? The physician has to take into account: (1) the needs of the patient (who wants assisted suicide); (2) the view of the medical community expressed through their ethical code (which prohibits assisted suicide); (3) the law of the land (which prohibits assisted suicide); and (4) a substantial body of medical and public opinion which claims the medical code and the law are outdated and should be changed (to facilitate assisted suicide). Whatever the physician chooses to do, reason can be used to argue the morality, or immorality, of the decision.

But there are limits to what the community can accept. When we reach this level of unequivocally immoral behavior, we are also dealing with unequivocal irrationality because what is bad for the community will also be bad for the individual. What if an individual rejects morality; that is to say, how do psychopaths fit in (or fail to fit in) to a culture? Blakiston's *New Gould Medical Dictionary* defines "psychopath" as "a morally irresponsible person." The notion of psychopathic immorality persists, although the term "psychopath" has

---

* "Against Lying," in *Treatises on Various Subjects,* ed. J. Deferrari (New York: Catholic University of America Press, 1952).

† Cited by Sissela Bok *(7).* The quotation comes from a letter published in *Briefwechsel Landgraf Phillips des Grossmüthigen von Hessen mit Bucer,* ed. Max Lenz.

been successively replaced, in psychiatric nomenclature, by "socio-pathic personality," "personality disorder, antisocial type," and "anti-social personality disorder" *(8)*. This evolution of psychiatric jargon reveals the increasing recognition that immorality is fundamentally an antisocial phenomenon, but since "antisocial personality disorder" is an unnecessary mouthful, we shall stay with the more traditional term "psychopath."

Psychopaths flout the universal, rational, ethical edicts on how people should live together. In his recent book *Without Conscience,* Robert Hare cites examples of psychopathic murderers. Three are enough to set the tone:

John Gacy, a Des Plaines, Illinois, contractor and Junior Chamber of Commerce "Man of the Year" who entertained children as "Pogo the Clown," had his picture taken with President Carter's wife, Rosalynn, and murdered thirty-two young men in the 1970s, burying most of the bodies in the crawl space under his house.

Charles Sobhraj, a French citizen born in Saigon who was described by his father as a "destructor," became an international confidence man, smuggler, gambler, and murderer who left a trail of empty wallets, bewildered women, drugged tourists and dead bodies across much of Southeast Asia in the 1970s.

Clifford Olson, a Canadian serial murderer who persuaded the government to pay him $100,000 to show the authorities where he buried his young victims, does everything he can to remain in the spotlight *(9)*.

Psychopaths lie as a matter of course: "The psychopath shows a remarkable disregard for truth and is to be trusted no more in his accounts of the past as in his promises of the future or his statement of present intentions. He gives the impression that he is incapable of ever attaining realistic comprehension of an attitude in other people that causes them to value truth" *(8)*. The sexual behavior of psychopaths is equally antisocial. Cleckley recalls: "I have seen psychopaths who seriously attempted to seduce sisters, mothers-in-law, and even their actual mothers. One boasted to his wife in glowing details of his erotic feats with her mother and his own" *(8)*.

A vivid picture of what it means to have no personal sense or morality—no conscience—is conjured up by this verbatim account quoted by Robert Hare. A psychopath describes how he murdered an old man during a burglary (and here the evidence indicated that he was telling the truth):

> I was rummaging around when this old geezer comes down stairs and . . . uh . . . he starts yelling and having a fucking fit so I pop him one in the, uh head and he still doesn't shut up. I give him a chop to the throat and he . . . like . . . staggers back and falls on the floor. He's gurgling and making sounds like a stuck pig [laughs] and he's really getting on my fucking nerves so I . . . uh . . . boot him a few times in the head. That shut him up. . . . I'm pretty tired by now, so I grab a few beers from the fridge and turn on the TV and fall asleep *(9)*.

From all these accounts psychopaths emerge as the most extreme noncooperators in society. Their actions are directed to immediate gratification of selfish desires, with total disregard for others. They prey on the community. When outrageous behavior is detected, the normal response of shame is missing. Equally important is the way psychopaths can use reason for their own purposes. Hare cites testimony from a psychopath after his twenty-third arrest for breaking and entering: "Sure I stole the stuff. But, hey! those folks were insured up to the kazoo—nobody got hurt, nobody suffered. What's the big deal?" *(9)*. So while reason is a tool for those who support morality, it can also be turned against them.

Psychopaths usually have excellent skills in communication; they have no disorder of thought, mood or memory. They do not experience anxiety, hallucinations, or delusions. They simply cannot grasp what is meant by words such as responsibility or obligation. This lack of conscience is reminiscent of the social inadequacies recorded in the celebrated case of Phineas Gage after his frontal brain injury. It is as if psychopaths have selective congenital damage to the module of brain function responsible for cooperation. Hare points out that "the premise of most correctional programs—that offenders have somehow gone off track and have only to be resocialized—is faulty when applied to psychopaths. From society's perspective, psychopaths have never been on track; they dance to their own tune" *(9)*. Attempts to treat psychopaths in therapeutic community programs have been unsuccessful; we have no answer to the problem. Serial killers and rapists are the tip of the iceberg. Psychopaths are responsible for some 50 percent of violent crimes and they account for about 20 percent of the criminals in prison. These figures may appear alarming, but it is even more distressing to know that only a small minority of psychopaths are in jail; most are at large *(9)*.

All criminals use reason to plead special mitigating circumstances—poverty, broken homes, religious fanaticism, or radical political ideology—and, of course, society uses reason to argue that people are responsible for their actions and the community must be protected from criminals. Reason is a tool for both sides. It is morally neutral in spite of the fact that it has played such a central role in the development of ethical values.

SOCIETY CONTINUES TO pose new moral questions and reason continues to help us find answers. As the end of the twentieth century approaches, ethical issues have gained a high profile. Attitudes to old questions like the death penalty go through cyclical reversals. With changes in circumstances, the weight of opinion may move in one direction, replacing controversy with consensus. No moral system will provide complete answers to all problems—there will always be disagreements within society. In tracing the rational arguments underlying morality we can detect a trend toward more individual freedoms and more general cooperation; it is easy to interpret this as progress toward a better society. But an interpretation more in keeping with the broader picture of evolution is that the changes are simply adaptations. Moral progress from this perspective is an illusory proposition. People are always responding, as best they can, to new situations, as their ancestors have done before. Reason has underpinned morality by adjusting the balance between cooperation and individualism, to meet the needs of the community in changing circumstances. More individual freedoms and more general cooperation become a formula for social stability as populations grow larger and accumulate different cultural groups.

6

COMMERCE

Alle thinges obeyen to moneye.
—GEOFFREY CHAUCER*

COMMERCE HAS AN interdependent relationship with reason. Commercial principles, such as "buy low and sell high," are inherently rational. Profit is the goal, and reason provides the way to reach it. To say that reason also depends on commerce seems odd, but if we look at history, commerce is the force that enabled reason to reassert its influence at the end of the European Middle Ages. Commerce brought power to people who had previously not been able to think or act independently. The new merchant classes can be said to have driven the Renaissance, the Enlightenment, and the Industrial Revolution. The merchants did not provide the intellectual power themselves, but they supplied the motivation by becoming patrons of science and art. Science helped merchants to succeed in their business enterprises, whereas art allowed them to celebrate and display

---

* Geoffrey Chaucer began work as a page to the Duke of Clarence, and was later transferred to the household of King Edward III. He came close to arrest for his debts, but for posterity he established the southern dialect as the language of English literature. Chaucer's father was the son of a tavern keeper, who was also deputy butler to Edward III. The quotation comes from "The Tale of Melibeus" in the *Canterbury Tales*.

their success. Because of this special relationship between reason and commerce, we shall explore how commerce arose and how it operates.

THE ORIGINS OF COMMERCE

Owning and trading are at the core of commerce. Can they be found, in an embryonic form, in the behavior of animals? What evidence would enable us to argue that animals "own," let alone "trade"? We can get clues from our nearest relatives, apes and monkeys. Several species seem to be possessive over food, their young, their mates, and their territory—a list with obvious relevance to survival of the species, so it is easy to see how it gained biological priority.

How do we infer that an object is the property of an animal? Frans de Waal has described what happens when food suddenly becomes available to a colony of chimpanzees: they throw themselves into each other's arms with obvious delight (1). If the supply is plentiful, every member of the colony gets a share. If the supply is limited, the socially dominant animals do not take everything: "It is not unusual to see the most feared and respected individual stretch out a hand to one of his underlings to beg for a scrap. Why does he not claim the food by force? In our tests, subordinates avoid dominants if both approach unclaimed food, but once the food is firmly in a subordinate's hands, his or her ownership is generally respected" (1). Property rights are the starting points of commerce. Equipped with language and reason, *Homo sapiens* has extended the rights to almost everything.

Can we find animal behavior that could represent the first steps toward trading? A trade is an exchange in which both parties benefit, so extortion doesn't count. Among chimpanzees, food is not passed from one animal to another out of fear; in de Waal's studies, dominant animals are, if anything, more likely to give than subordinates. The principle underlying the transfer of food appears to be *reciprocity*. De Waal continues: "As predicted by the reciprocity hypothesis, the number of transfers in each direction was related to the number in the opposite direction" *(1)*. If Alan shared generously with Betty, Betty generally shared generously with Alan, and if Alan shared little with Charlie, Charlie also shared little with Alan. This reciprocity hypothesis is further supported by the finding that grooming affected the sharing of food; Alan's chances for getting food from Betty improved if Alan had groomed her earlier that day. Once a *quid pro quo* mindset had taken hold, the "currency" of exchange became secondary, and reciprocity seemed to permeate all aspects of social life.

Homo sapiens has built upon the principle of reciprocal altruism to create contracts. The essence of reciprocal altruism is the *likelihood* that a favor will be returned: I will do something for you, and at some future time you will *probably* do something for me. The essence of a contract is the *obligation* that a favor will be returned and the defining of each favor: I *shall* do this for you, and in return you *shall* do *that* for me. Language and reason allow the move from probable to certain, and from general to specific, with punishment if the contract is broken.

The simplest form of contractual trading is barter, but this relies upon two people finding each other with the right goods to give in exchange—a chancy business. Reason solved the problem by separating the barter transaction into buying and selling, through the creation of a symbol of value—money.

MONEY

Money is a product of reason. Although money would have started as an instrument for facilitating the exchange of goods and services, it brought other benefits. It offered a convenient way to collect and hold taxes, and it allowed individuals and communities to build savings. A surplus of money is more versatile and therefore more convenient and useful than a surplus of goods. John Stuart Mill described money as "a machinery for doing quickly and com-

modiously, what would be done, though less quickly and com-
modiously, without it and like many other kinds of machinery, it
only exerts a distinct and independent influence when it gets out of
order"* (2).

Since money is merely a symbol of value, it can take many forms,
but shifting between currencies may be tricky. A. Hingston Quiggin
describes the experiences of a nineteenth-century traveler in Africa:

> He, like Stanley, started off with the usual assortment of trade
> beads and trade goods, but found, on crossing Lake Tanganyika,
> that the beads were no longer currency, and he had to lay in a stock
> of copper crosses. When he reached Nyangwe on the Lualaba, an
> important market town, where he was expecting to buy a canoe
> and continue his journey, nothing was accepted except slaves,
> goats and cowries (3).

What is the most rational form of money?

> The ideal properties of money are that it shall be handy, lasting,
> easy to count, and difficult to counterfeit (portable, durable, di-
> vidable and recognizable), so there are not many rivals to the pre-
> cious metals. The most remarkable exception is the cowrie, which,
> starting on its travels before gold and silver coinage was in use, ex-
> tended its range further than any other form of money before or
> since, spreading from China and India eastward to the Pacific Is-
> lands; traveling across and encircling Africa to the West Coast; and
> penetrating into the New World (3).

---

* John Stuart Mill had an unusual education. His father was a Scottish philosopher who
set a rigorous intellectual pace for his son. At the age of three years he began learning
Greek; at eight years, Latin and arithmetic; at twelve years, logic; at thirteen years, political economy. He became the leading exponent of utilitarian moral philosophy. He fought
hard for women's rights, publishing *The Subjection of Women* in 1869. He is best known
for his essay *On Liberty* (1859) and his book *Principles of Political Economy* (1848). He was
the godfather of Bertrand Russell.

Looking back into preliterate history, it is difficult to be sure what was money. When an object and a currency unit had the same name, the object was probably used as money. Small items found in large numbers, such as beads and shells, were also probably money. Copper, bronze, and brass rings seem to have been used as money some 6,000 years ago. The earliest real coins were found in Turkey, dating from 2,700 years ago. While metal coins and printed paper are now the normal forms of currency, over the twentieth century many other materials have been employed in different parts of the world— large stone discs, feathers, beeswax, dogs' teeth, whales' teeth, boars' tusks, and still, as previously, cowrie shells *(4)*.

THE BIBLE TELLS us that "a feast is made for laughter, and wine maketh merry; but money answereth all things" (Ecclesiastes 10:19). George Bernard Shaw agreed: "Money is, indeed, the most important thing in the world; and all sound and successful personal and national morality should have this fact for a basis"* *(5)*.

Elsewhere, Shaw goes further: "Food, clothing, firing [heating], rent, taxes, respectability and children. Nothing can lift those seven millstones from man's neck but money; and the spirit cannot soar until the millstones are lifted" *(6)*. For most people, in most parts of the world, money is simply a means to survival, and, as Shaw points out, a vitally important means.

---

* George Bernard Shaw was born in Dublin in 1856, and his early life was not easy. Initially he worked in a land agent's office. When he came to London his first five novels were rejected by the major publishers. In 1925 he received the Nobel Prize for literature. Shaw was known as a devout socialist, a member of the executive committee of the Fabian Society, and an editor of *Fabian Essays*. Yet he made some incongruous political statements, as pointed out by John Carey in his book *The Intellectuals and the Masses* (1992). Shaw referred to the democratic electorate as "the promiscuously bred masses," and he declared that "the majority of men in Europe at present have no business to be alive." In the Preface to *On the Rocks* Shaw asserts: "Extermination must be put on a scientific basis if it is ever to be carried out humanely and apologetically as well as thoroughly. . . . if we desire a certain type of civilization and culture, we must exterminate the sort of people who do not fit in with it."

Money is as important for communities as it is for individuals. Historically, the military success of nation-states has been determined by their economies. In *The Rise and Fall of the Great Powers,* Paul Kennedy argues that while skill, folly, or luck may seal the fate of specific battles, economic factors have always sealed the fate of large-scale wars: "In a long-drawn-out Great Power (and usually coalition) war, victory has repeatedly gone to the side with the more flourishing productive base—or, as the Spanish captains used to say, to him who has the last escudo" *(7).*

TRADE

Long-distance commerce can be traced back five thousand years to trade between Egypt and Mesopotamia. In ancient times the Phoenicians, the Greeks, and the Romans traded widely—they were, collectively, the founders of modern commerce. "The birth of the economic revolution of the Middle Ages took place, not in the static agrarian feudal society of Western Europe, but rather in the dynamism of trade and industry inherent in most of the countries of the Eastern Mediterranean" *(8).* From the fifth to the seventh centuries, the Byzantine Empire became the center of trade.

From the eighth to the eleventh centuries the Arabs and Persians were the dominant traders; then the action moved to Western Europe as Belgium and Holland became commercial hubs, probably because of their geography. Their long coastlines and numerous waterways gave access and protection to shipping and their location allowed them to forge links with the overland caravan routes. The Low Countries were a nodal point for trade moving to and from the East, Italy, and England *(9).* European commerce started to flourish at a time when castles were being built on strategic routes for defense, which were also the trading routes.

Early merchants were adventurers, often traveling long distances to seek their fortunes; their independence contrasted with the rest of feudal society, but they were tolerated because they brought what was wanted. They discovered, in the words of N. Rosenberg and L. E. Birdzell:

> the enormous rewards to be reaped from introducing a new product that was popular with buyers and had no immediate competitors. They may have scandalized their late-medieval colleagues by skimming off consumers' money for exotic foreign goods instead of the sober products of the local guilds, and they may have outraged their fellow burghers by drawing promising youngsters from honest trades into the hazards of voyages to unknown and often pagan places. But in modern terms, what they did is called innovation *(10)*.

This innovation, which brought wealth to the Western nation-states, was only made possible by the decentralization of political control in Europe after the fall of the Roman Empire. In contrast, the monolithic control of imperial China sustained a rigid society lacking the freewheeling flexibility necessary for commercial expansion. Knowledge and skill were, in many respects, more developed in China than Europe, but the universality of Chinese imperial rule discouraged the individualism that was to allow European commerce to grow and culminate in long sea voyages of mercentile discovery.

THE CHURCH FROWNED upon financial gain.

> In the Middle Ages the Church taught that "No Christian ought to be a merchant," and behind that dictum lay the thought that merchants were a disturbing yeast in the leaven of society. In Shakespeare's time the object of life for the ordinary citizen, for everybody, in fact, except the gentility, was not to advance his station in life, but to maintain it.... The early capitalists were not the pillars of society, but often its outcasts *(11)*.

Yet the church had an obvious conflict of interest in a situation where personal gain would erode its power, for a "traditional economy" had given the church substantial assets. The critical need for the development of commerce was the ability to borrow large quantities of money. The problem was solved through the rational recognition that money was like other forms of property, so it could be rented. For this to become widespread, a new kind of merchant was required—the banker. But the principle of renting money—making a profit from a loan—was forbidden by the church. Nevertheless, people soon found a way to evade the ban. Loans were commonly repaid in another currency at a "special" exchange rate that concealed the profit. In medieval Europe, Jews were prevented from owning land and excluded from most guilds, so they pursued banking as one of the few permitted activities. The concepts of loans and interest were old, but banks legitimized and promoted them. Written promises and claims for money became widely accepted and checks began to be written in the seventeenth century. Banks made a profit by charging commissions for their services, but more importantly they lent at a higher rate of interest than the interest they paid on the deposits that they received. A bank's main asset was the money owed to it.* Central banks were established as national institutions responsible for controlling the supply of money, the cost of credit, and the foreign exchange rate. The Bank of England was incorporated by an act of Parliament in 1694.

In spite of ecclesiastical discouragement in the Middle Ages, commerce steadily grew because people obviously benefitted from it. Trade brought wealth and fostered competition. In a free market, commercial success goes to those who give the best value, so trade is a rational instrument to stimulate more efficient production of better articles and commodities. In due course merchants and bankers demanded a level of independence that was to have far-reaching po-

---

* Everyone assumes their loans will be repaid, but major defaults can occur, leading to serious financial difficulties—as happened in the 1990s in Asia.

litical consequences. "The change from the coherent, fully integrated feudal society of the late Middle Ages to the plural society of eighteenth-century Europe implied a relaxation of political and ecclesiastical control of all spheres of life" *(10)*. The "relaxation" was a rational response to the fact that when the merchants and bankers built up wealth, they built up power. They may have been disdained, but they could not be ignored.

Economics

Economic theories rest upon observation, rational argument, and mathematical analysis—but economics cannot employ rigorous scientific methods because hypotheses cannot be tested by controlled experiments. In contrast, for example, medical theories about new treatment can be investigated by conducting comparisons in which neither the physician nor the patient know which of two alternative therapies is being used. This design eliminates bias, but we still demand replication of the experiment. If the hypothesis survives all these tests, it is accepted. Economic theories cannot be tested in this way.

ONLY THREE TYPES of economic system have been found to work. *Traditional economies* have operated where the community is stable and rigidly stratified—for example in feudal Europe and in India under the caste system. *Command economies* have operated when governments exercise firm, centralized control—monarchies and dictatorships. Megaprojects, such as the building of the Egyptian pyramids and the Great Wall of China, were financed in societies that expected, and accepted, coercive authority. *Market economies* are more recent.

Ideally, in a market economy large numbers of producers use labor, knowledge (technology), capital, and natural resources to supply everything the community demands. Each producer adjusts output to seek a maximum profit. When all producers do this, supply changes and there is an automatic movement of prices. Too much production leads to a fall in prices and too little leads to a rise. Competitive pricing sets an equilibrium between supply and demand, and balances the interests of producers and consumers. Everything is driven by the profit motive of the producers, but competition protects the consumers from exploitation. Monopolies undermine the equilibrium, but they may be necessary, in a regulated form, to provide essential services.

All of this is highly rational, though oversimplified. Adam Smith called the process of automatic equilibration the "invisible hand" controlling the free market. He set out a carefully argued blueprint for industrial expansion in *The Wealth of Nations** (12). This landmark book was published in 1776, as the Industrial Revolution was in its infancy, with steam engines being introduced to pump water out of coal mines. The real power of Smith's book is the suggestion that because commerce is automatically controlled by the "invisible hand," an ideal free-market economy can operate independently and benefit everyone because self-interest is curbed by competition. Unfortunately, free-market economies in the real world are not ideal—they can encounter serious problems, such as prolonged unemployment *(13)*.

But when it was published, the optimistic analysis of the free market economy by Adam Smith was received with enthusiasm, for it seemed to offer a rational explanation and justification for the economic changes under way as the Industrial Revolution gathered momentum.

THE ROOTS OF the Industrial Revolution lay in the rational effort to increase efficiency in production. Paul Kennedy describes how this was achieved:

> What industrialization, and in particular the steam engine, did was to substitute inanimate sources of power; by converting heat

---

* Adam Smith was the son of a Scottish comptroller of customs. He became professor of logic at Glasgow, and later professor of moral philosophy. Smith moved in the same social circle as David Hume, Samuel Johnson, Joshua Reynolds, and David Garrick.

into work by the use of machines—rapid, regular, precise, tireless machines—mankind was thus able to exploit vast new sources of energy. The consequences of using this novel machinery were simply stupendous: by the 1820s someone operating several power driven looms could produce twenty times the output of a hand worker, while a power-driven mule (or spinning machine) had two hundred times the capacity of a spinning wheel. A single railway engine could transport goods which would have required hundreds of pack-horses, and do it far more quickly *(7)*.

Private enterprise and industrialization, operating in a free-market economy, led to enormous gains in wealth. In fact, aggressive, uncontrolled pursuit of money was too successful. It strained the morality that held society together. Trade unions were a rational response to exploitation of workers, and strikes showed how unbridled self-interest by the few could provoke massive noncooperation by the many. The strikes brought harsh retaliation, but they also drew attention to squalid working conditions. The workers' cause was gradually taken up by influential intellectuals such as Émile Zola.* Social reformers effectively pointed at the disgrace of widespread industrial practices such as child labor:

> The golf links lie so near the mill
> That almost every day
> The laboring children can look out
> And see the men at play *(14)*.

In these circumstances, governments were forced to look more critically at the consequences of unbridled commerce operating in a free-market economy, and they used reason to find practical solu-

---

* Zola was deeply moved by the coal field strikes at Anzin, in northeastern France. In order to understand the issues, he spent several months visiting the mines, the miners' cottages, and the beer halls where the miners met. His epic *Germinal* was published in 1885.

tions. Ethical principles for communities are similar to ethical principles for individuals—and for communities, as for individuals, solutions are good if they promote survival. The expansion of capitalism presented governments with two interrelated problems: the poor were demanding a redistribution of the wealth to create "social justice," and the rich were becoming fearful that continuing repression of the workers would lead to revolution. The challenge was, and is, to find a balance. The fact that some people have more and others less can lead to social instability if the distribution is unacceptable to the prevailing view of what is right—but the prevailing view of what is right varies from time to time and place to place. Governments have to maintain a continuing balance between, on the one hand, the need to help the poor, and on the other hand, the danger of killing the profit motive that leads business people to work hard, innovate, and take financial risks when necessary. As a reasoned response, governments increased taxes to fund welfare programs, and regulated business practices and investment; they extended education to all, and they passed laws against discrimination, thus allowing anyone to enter the world of commerce. Naturally governments can never satisfy all opinions on how to distribute wealth fairly and sustain the economy. A society that is adapting optimally to new social and economic situations has to make repeated adjustments.

CONTROVERSIES SURROUND ECONOMIC assertions because they so often have political overtones. Staunch advocates of capitalism will deny that their system leads to injustice, and staunch advocates of socialism will deny that their system leads to inefficiency. Both sides find evidence to support their positions, and both sides find dismissive language to deal with the opposition. For example, Noam Chomsky questions the widely held view that egalitarianism results in loss of efficiency. "In general the relation between equality and efficiency is hardly a well-established one, despite many facile pronouncements on the matter" (16). But economic evidence, and its interpretation, are biased by right and left political wings alike and "facile pronouncements" abound. Perhaps the most compelling example of socialism getting into economic difficulties is the collapse of the Soviet Union during the 1980s. The experience of two centuries suggests that socialism without capitalism leads to inefficiency, and capitalism without socialism leads to exploitation. When socialism and capitalism are brought together, they support each other— the term "mixed economy" reflects this interdependence.

Economic theories apply reason to such issues as unemployment, the supply of money, public spending, and international trade. They deal with "how men and women obtain their livelihoods" *(17)*. Judgments that shape the distribution of wealth must be driven by moral values, so economists become involved in deciding what is good and what is bad. J. K. Galbraith has recently spelled out his vision of an economic and ethical ideal:

> Just what should the good society be? Towards what, stated as clearly as may be possible, should we aim? The tragic gap between the fortunate and the needful having been recognized, how, in a practical way, can it be closed? How can economic policy contribute to this end? What of the public services of the state; how can they be made more equitably and efficiently available? How can the environment, present and future, be protected? What of immigration, migration and migrants? What of the military power? What is the responsibility of the good society as regards its trading partners and neighbors in an increasingly internationalized world and as regards the poor of the planet? *(18)*

Galbraith gives a general answer to his questions:

> Let there be a coalition of the concerned and the compassionate and those now outside the political system, and for the good society there would be a bright and wholly practical prospect. The affluent would still be affluent, the comfortable still comfortable,

but the poor would be part of the political system. Their needs would be heard, as would the other goals of the good society. Aspirants for public office would listen. The votes would be there and would be pursued. And now with the safety net, health care, the environment and especially the military power, the good society fails when democracy fails. With true democracy, the good society would succeed, would even have an aspect of inevitability *(18)*.

Galbraith sets out goals which he claims are attainable with the help of reason. Of course, no one can argue with Galbraith's expression of human decency. In 1759, Adam Smith wrote: "How selfish soever man may be supposed, there are evidently some principles in his nature, which interest him in the fortune of others, and render their happiness necessary to him" *(19)*. But as Frank Hahn points out, the issues are difficult: "The interesting question is not 'Do we care about others?' but 'How much do we care and for which others?' " *(19)*.

More questions arise from Galbraith's impassioned plea for a society where the economy is controlled to achieve an ideal way of life. Who defines utopia in a democracy? Each political party has its own version, but it is a universal feature of democracy that governing parties change. More plausibly Galbraith's good society could simply be rational adaptation to the social pressures under which we live. This interpretation of the good society is in keeping with the general biological and evolutionary theme of our earlier arguments. Our society includes many different kinds of people, and some have more education and more financial power than ever before. Political and economic systems have to adjust to these realities. One rational, and arguably the most rational, response is the creation of something very like Galbraith's vision of the good society. This analysis supports Galbraith's conclusions, but assigns to reason the attainable task of optimizing adjustment rather than the more ambitious task of designing a utopia. In this view, commerce is not an engine of *progress* toward some better social or material world. It is an engine for *adaptation*, helping us to make adjustments as an increasingly large and diversified population seeks wealth and continues to accumulate knowledge.

EVEN THOUGH WE are not heading for a utopia, we can still look into the future with hope as we seek rational ways of dealing with new circumstances. One contemporary adjustment is increasing economic interdependence between nations. As the corporations of the developed countries become multinational, the most powerful nation-states share technology, manufacturing plants, suppliers of raw materials, and markets. Such internationally enmeshed commercial interests tend to make war a "no-win" economic option. Europe offers an example of these changes in action: over the last millennium the various European nations have been at war with each other or they have been slowly recovering from the previous war and getting ready for the next. Wars are the milestones of European history, yet this tradition is now fading as the nations of the European Union move toward political federation.

We can also see the increasing commercial power of knowledge. The Enlightenment and the rise of science were driven by freedom generated through commerce—freedom from subservience to the monarch and the church. Thus commerce, reason, and science have always been linked, but now the ties are stronger, through the increasing power of increasing knowledge.

The basic economic resource—the means of production, to use the economist's term—is no longer capital, nor natural resources (the economists' land), nor labor. *It is and will be knowledge.* The central wealth-creating activities will be neither the allocation of capital to productive uses, nor labor, the two poles of nineteenth- and twentieth-century economic theory, whether classical, Marxist, Keynesian, or neo-classical. Value is now created by productivity and innovation, both of which are applications of knowledge *(15)*.

But there is another side to the coin. We have to ask, in the words of Paul Kennedy:

> whether today's global forces for change are not moving us beyond our traditional guidelines into a remarkable new set of circumstances—one in which human social organizations may be unequal to the challenges posed by overpopulation, environmental damage, and technology-driven revolutions and where the issue of winners and losers may to some degree become irrelevant. If, for example, the continued abuse of the developing world's environment leads to global warming, or if there is a massive flood of economic refugees from the poorer to the richer parts of the world, everyone will suffer. In sum, just as nation-state rivalries are being overtaken by bigger issues, we may have to think about the future on a far broader scale than has characterized thinking about international politics in the past. Even if the Great Powers still seek to rise, or at least not to fall, their endeavors could well occur in a world so damaged as to render much of that effort pointless *(21)*.

COMMERCIAL MENTALITY

People who succeed in commerce have a predisposition for taking action and taking risks when necessary. As John Maynard Keynes put it: "If human nature felt no temptation to take a chance, no satisfaction (profit apart) in constructing a factory, a railway, a mine, or a farm, there might not be much investment merely as a result of cold calculation" *(13)*. Commerce is rational, but it is driven by our irrational emotions.

Among business people there is a special breed who have exceptional skills at getting things done, and are prepared to take exceptional risks—the entrepreneurs. Joseph Schumpeter illustrates how they operate.

The function of entrepreneurs is to reform or revolutionize the pattern of production by exploiting an invention or, more generally, an untried technological possibility for producing a new commodity or producing an old one in a new way, by opening up a new source of supply of materials, or a new outlet for products, by reorganizing an industry and so on. Railroad construction in its earlier stages, electrical power production before the First World War, steam and steel, the motor car, colonial ventures afford spectacular instances of a large genus which comprises innumerable humbler ones—down to such things as making a success of a particular kind of sausage or toothbrush. This kind of activity is primarily responsible for recurrent "prosperities" that revolutionize the economic organism *(21)*.

Entrepreneurs are dedicated to innovation, and the single-minded pursuit of their visions overrides all other considerations, including risk.

THE LARGEST GROUP of people involved in the economy are those who have to spend everything they earn to buy whatever they need most. The choice of "what they need most" is usually emotional. The next-largest group are the millions of small investors in the world's stock markets. Investors develop rational techniques to profit from the stock market. They use formulae to estimate the value of stocks; they employ mathematical models to analyze the cyclical fluctuations in prices; they devise indices to make early predictions of new market trends; they even study the social psychology of other investors. Reason plays a prominent part in this huge industry, but the variables are so numerous and complex that all too often financial storms break without warning, and the devastation may extend far beyond the world of the investors. One factor contributing to the instability of the system is the irrational but decisive influence of emotions. Charles Kindleberger describes these unruly forces in his book *Manias, Panics and Crashes.*

Here are some phrases culled from the literature: *manias... insane land speculation ... blind passion ... financial orgies... frenzies... feverish speculation . . . epidemic desire to become rich quick . . . wishful thinking . . . intoxicated investors . . . turning a blind eye . . . people without ears to hear or eyes to see . . . investors living in a fool's paradise . . . easy credibility . . . overconfidence . . . overspeculation . . . overtrading . . . a raging appetite . . . a craze . . . a mad rush to expand (23).*

Emotions and cultural attitudes play an important part in another area of the economy—day-to-day business decisions. In contrast to the disruptive influence of emotional forces on the stock market, Robert Frank argues that emotions have a useful part to play in business relationships. Emotional reactions are less predictable than reason, and a negotiator who is known to be unpredictable has an advantage. Frank also argues that a sound business relationship is more likely to be engendered by warm emotion than cool reason. In wider perspective, business people who are helpful to their friends will receive help back, and in the long term they are more likely to do well. *(24)*

This conclusion is not in keeping with the traditional economic concept of rational behavior, described by Jon Elster: "In the older body of literature rational-choice models were often associated with the assumption that behaviour is motivated merely by egoistic, hedonistic, or narrowly self-interested purposes, and the phenomena of altruistic behaviour were either denied or believed to create an insuperable anomaly" *(25)*. But the disadvantages of ruthless, short-term self-interest are revealed by repeatedly playing out the "prisoner's dilemma."* The old idea that reason dictates selfishness is simply untenable. Robert Axelrod has explored the outcome of different strategies when we do not know other people's intentions. He reached the unexpected conclusion that "the key to doing well lies

---

* See chapter 13, page 289.

not in overcoming others, but in eliciting their cooperation" *(26)*. Richard Dawkins has argued the same case in *Nice Guys Finish First (27)*. Through short-term, aggressive competition, we have hunted many species to the verge of extinction, and we have exhausted rich fishing grounds; these and many similar disasters could have been prevented by cooperative, long-range planning. It seems odd that short-term self-interest has dominated commerce so often, for our society is built upon cooperative long-term understandings and agreements that are the moral backbone of our culture. Where is the optimal balance between competition and cooperation, and how do we attain it? There do not seem to be any simple answers, for the balance changes with circumstances. When a community is threatened, people cooperate more, and when the threat passes, they compete more. In the long term, competition and cooperation are both necessary and both are rational. They both have a high profile in the world of commerce, but they are both to be found in virtually every walk of life. Competition and cooperation have to be balanced just as we reconcile other ethical conflicts, and the new evidence on the long-term benefits of cooperation may influence how we reset the balance in the future.

So an element of trust is beneficial to everyone and the notion that reason only dictates short-term selfish action is patently wrong. We have defined reason as an efficient way of reaching goals, and it follows that commercial cooperation is rational. People who make fortunes are neither more nor less happy than those who are just financially comfortable. Why, then, do the "high flyers" of business continue to work when they already have all the money they need? Of course, we have to consider what we mean by "all the money they need." People have different standards of what constitutes an adequate life style, but when wealth reaches billions of dollars, we must look beyond needs. Highly successful business people continue to work because they enjoy what they do best. In addition, there is social motivation—the trappings of power, status, and respect. Another social motive of importance comes from the microculture of commerce in which "winning" plays a big part. There are different leagues for commerce just as there are different leagues for sport. In commerce as in sport, the contestants in each league jockey to reach the top, because their microcultures motivate them to do so. The stars in commerce and sport are driven by their ambitions to outperform the rest. In the words of Red Sanders, coach of the UCLA football team: "Winning isn't everything, it's the only thing" *(28)*.

7

———

GOVERNMENT

Law in general is human reason.
—BARON DE MONTESQUIEU*

THE PRIMARY RESPONSIBILITY for government is to trans-
late ethical principles into practice by creating, interpreting, and enforcing laws. Reason is an essential instrument for turning morality
into reality, and morality is simply a rational formula for living together—so reason is entrenched in both the purpose and operation
of government. The most challenging responsibility for government
lies beyond the central, universally accepted moral standards in the
gray zone where individuals disagree, sometimes violently. These are
the controversial general ethical dilemmas of public concern—such
as how to distribute the tax burden and where to spend public
money.

In addition to the moral questions that face government, there are
the practical matters of who should govern and how. David Hume
marvelled at how any government can work: "Nothing appears more
surprising to those who consider human affairs with a philosophical

---

* *Défense de l'Esprit des lois,* published in 1750. Montesquieu was a magistrate, a jurist,
and a scientist. His major impact on the world, however, was his powerful advocacy of
democracy as a political philosophy. His influence was felt in both Europe and North
America, for his ideas contributed to revolutions on both continents.

eye, than the easiness with which the many are governed by the few."*
While governmental leadership may be easier than expected, it is still
difficult to assign priorities to objectives such as prosperity, social
justice, security, prestige, and influence. Decisions will depend on
who the rulers are, or more precisely, what the goals of the rulers
happen to be—and the goals of the rulers are determined by their
particular subcultures. Attitudes set by the culture are so deeply
rooted in policy that, in a sense, the government functions as the executive of the culture.

Since governments must have power in order to operate, communities have to settle how they will grant power, how they will regulate
it, and how they will transfer it. History records a trend in the way
cultures have approached these tasks, from family elders to tribal
chiefs, from tribal chiefs to kings, and from kings to democracies;
each step was a rational response to the increasing complexity and
power of the population to be governed. The trend may be seen as
teleologically orchestrated, long-term progress or alternatively as a
series of short-term adjustments to changing circumstances, with no
preordained direction. The biological evidence from evolution supports the latter interpretation if we are looking for the simplest explanation for how we came to be where we are—and according to
Ockham's razor, the simplest explanation is the most rational.

ELDERS, CHIEFS, AND KINGS

The earliest human communities were small, and they must have
been organized very simply. Their social structure can be inferred
from currently surviving isolated groups of people in the Amazon
and upper Nile regions, where tribes and villages are broken up into

---

* From Hume's "Of the First Principles of Government" (1741), in *Essays Moral, Political
and Literary,* ed. E. F. Miller (Indianapolis: Library Classics, 1987). Hume's life was not
easy. He studied at Edinburgh University but did not graduate; later he was blocked, because of his atheism, from the professorship of moral philosophy in Edinburgh and the
professorship of logic in Glasgow. He had bouts of depression, and at one stage he worked
as a counting-house clerk in Bristol.

extended families. Often the family elders have an equal voice, but sometimes they identify a chieftain to whom they give limited power—for little power is necessary.

With the development of agriculture, 12,000 to 6,000 years ago, communities began to grow. The size of their populations increased to several hundred, and more ambitious cooperative projects could be undertaken, such as irrigation. As communities became larger, they had to become more structured to operate as a cohesive group. A leader became essential; he was usually chosen by the tribal elders. The selection process generally required magical signs or religious rituals—a small-scale version of the search for a new pope. Wars became a regular part of life, so potential chiefs were expected to be strong and courageous. When a chief's physical prowess or religious talent declined, a new one was sought through the same methods. Religion nourished the emotional bonds keeping people together, and reason helped to find social structures that would work.

These arrangements sufficed until communities grew to city-states, when the population became stratified to take on specialized roles. Then a new kind of government became necessary. Again, the social organization was rational, but on a bigger scale: give the leader more authority—by establishing a standing army—and ensure stability with a simple method of transferring power when it became necessary. In short, the city-states invented hereditary monarchy by applying reason to the task of managing a community too large and complex to be handled by tribal chiefs, yet too uneducated to operate as a democracy. Monarchy has been a resilient, enduring style of government from the dawn of literate history until very recently.

The early literature on monarchy is full of praise for the system—perhaps no surprise since dissidents were unlikely to survive. In the twelfth century, John of Salisbury wrote:

Therefore the prince stands on a pinnacle which is exalted and made splendid with all the great and high privileges which he deems necessary for himself. And rightly so, because nothing is more advantageous to the people than that the needs of the prince should be fully satisfied; since it is impossible that his will shall be found opposed to justice. Therefore, the prince is the public power, and a kind of likeness on earth to the divine majesty.*

This smacks of sycophancy, and indeed, most of the recorded comment on monarchy reflects situations in which admirers flourished and critics risked their lives.

Monarchs took responsibility for prosperity and social stability, defending the realm, and embarking upon conquests. They used reason to run a bureaucracy that would enact laws and impose taxation for the financial base upon which they, and everything else, depended. Governments had to take responsibility for the finances of the realm; this meant it had to control the supply of money.

These changes allowed large populations to be governed for long periods, but sooner or later difficulties arose because of social unrest during famine, pestilence, or unsuccessful war. Monarchs needed friends to help them through hard times, so they formed alliances with the priesthood and created an aristocracy of loyal secular supporters. If they were lucky, brave, and clever, monarchs survived and extended their domains. These were the circumstances under which monarchies continued to operate and reinforce their control on society. In the fourteenth century Dante wrote: "Who doubts now that a Monarch is most powerfully equipped for the exercise of Justice? None save he who understands not the significance of the word, for a Monarch can have no enemies. The assumed proposition being

---

* John of Salisbury was a prelate and ecclesiastical spokesman. He was secretary to Thomas à Becket. This quotation comes from *Polycraticus,* his treatise on church and state diplomacy. He was appointed bishop of Chartres in 1176.

therefore sufficiently explained, the conclusion is certain that Monarchy is essential for the best ordering of the world."*

Machiavelli applied reason to the task of producing a manual on how to rule in order to sustain power:

> From this arises the question whether it is better to be loved more than feared, or feared more than loved. The reply is, that one ought to be both feared and loved, but as it is difficult for the two to go together, it is much safer to be feared than loved, if one of the two has to be wanting. For it may be said of men in general that they are ungrateful, voluble, dissemblers, anxious to avoid danger, and covetous of gain; as long as you benefit them, they are entirely yours; they offer you their blood, their goods, their life, and their children, as I have before said, when the necessity is remote; but when it approaches, they revolt. . . . And men have less scruple in offending one who makes himself loved than one who makes himself feared; for love is held by a chain of obligation which, men being selfish, is broken whenever it serves their purpose; but fear is maintained by a dread of punishment which never fails *(1)*.

Whether all the observations on human nature are correct may be open to question, but the argument is certainly well reasoned.†

Jacques-Bénigne Bossuet (1627–1704) was an ecclesiastic who became tutor to the dauphin. He was an ardent enthusiast for the monarchy.

---

* Dante was the son of a Florentine lawyer. This quotation comes from *De Monarchia*, written about 1313, in which he proposes world government by a universal emperor. His writing was dominated by poetry, but he also had an interest in political theory and in the origin of language.

† Machiavelli knew what he was writing about from firsthand experience. In 1513 he was arrested on a charge of conspiracy, and he was tortured. Later he was pardoned, but he had to withdraw from public life. This is when he drafted what he called his "little book."

Men are all born subjects, and by nature are accustomed not only to obey, but also to having only one leader. Monarchy is the best system. It is the most natural, it is consequently the most durable and the strongest. It is also the most opposed to that division that is the worst evil of states and the most certain cause of their ruin. Of all forms of monarchy, the best is heredity where succession is in the male line and in order of seniority. . . . We have established by the Scriptures that royalty has its origin in divinity itself. God chose the monarchical and hereditary state as the most natural and the most durable.*

It is difficult to find a more resounding affirmation of the system. Bossuet was rational, even if his writing might be considered self-serving. It was irrefutable that hereditary monarchy had worked well in the conditions that had prevailed throughout literate history, for like a dominance hierarchy of apes or ants, the group survived because there was strength in numbers.

BUT THE WORLD was changing; less than a century after Bossuet's homage to princes, revolution shattered the French monarchy and sounded the death knell of dynastic power everywhere. Kingdoms were replaced by democratic republics or by constitutional monarchies in which royal families survived only in name, without the power to govern. How and why did such cataclysmic events occur? Did reason play a part in the fall of the monarchies? The seeds of unrest were sown by commerce, the Renaissance, the Reformation, and the printing press. Reason played a central role at each step, and each step brought power to people who had never had it before.

Hitherto, monarchies had collapsed if they were unlucky or exceptionally stupid, but under normal circumstances a failing monar-

---

* Bossuet, trained in a Jesuit school in Dijon, was one of the first to attempt a philosophy of history. He was a distinguished intellectual and churchman who preached to Louis XIV in 1661 and gave the funeral oration for Henrietta Maria in 1669. This quotation comes from *Politique tirée de l'écriture sainte*, published in 1709.

chy would simply undergo internal replacement or be vanquished by an external enemy. A new monarchy would arise from the ashes of the old or a new colony would be grafted onto an existing empire. Either way, the monarchistic system of government would not be jeopardized, for like a theatrical play, the action would continue with new actors taking the old parts. We must look further to see why the system of government that lasted so well for thousands of years underwent a virtually worldwide collapse over the course of a mere 500 years. Because erosion of the authority of the king of England was more gradual than the abrupt collapse of many other dynasties, English history illustrates, in slow motion, how monarchy eventually lost its grip.

We have seen that one of the essential components of a successful monarchy is a standing army, but armies are expensive, and a thousand years ago it was increasingly difficult to collect taxes in England. Heavily armed horsemen—knights—were considered to be the decisive weapon, but they had to finance themselves because their equipment was too expensive for the royal purse. The knights with the greatest wealth and influence built impregnable fortresses, and became relatively independent of their monarchs. The English warlords spent some two hundred years fighting each other, and eventually acquired sufficient influence to threaten their king. They collectively demanded a share of his power, and King John had no option but to concede; he entered into a formal agreement to decentralize political control of the country by signing the Magna Carta in 1215. Here reason revealed that political and military power had passed to the warlords to such an extent that King John could not retrieve it without risking a war that he would probably lose. In this situation, the only rational response was to sign.

The discontent among the aristocracy that led to the Magna Carta slowly spread to more humble members of the community whose tradition of allegiance and submission had derived from impotence. Through the printing press, introduced in the fifteenth century, more people were becoming informed, and at the same time, through commerce, they were gaining power. These changes had rational origins, for printing was a product of mechanical engineering, and commerce was the result of economic principles applied to costs, profits, loans, and control over the value of money. Increasing diffusion of both knowledge and financial power created widespread frustration with the monarchy. A parliament was established and in 1583 Sir Thomas Smith described how it worked; he was in a good position to comment, since he had been at various times professor of civil law and Greek in Cambridge, provost of Eton, dean of Carlisle, secretary of state, and ambassador to France.

> The most high and absolute power of the realme of Englande, consisteth in the Parliament. For as in warre where the king himselfe in person, the nobilitie, the rest of the gentilitie and the yeomanrie are, is the force and power of Englande; so in peace and the consultation where the Prince . . . the Baronie of the nobilitie and higher, the knightes, esquires, gentlemen and commons for the lower part of the common wealth, the bishops for the clergie bee present to advertise, consult and shew what is good and necessarie for the common wealth, and to consult together, and upon mature deliberation everie bill of lawe thrise reade and disputed uppon in either house . . . the Prince himself in presence of both the parties doeth consent unto and alloweth. That is the Princes and whole realmes deede: whereupon justlie no man can complaine, but must accommodate himself to find it good and obey it *(2)*.

Thus Sir Thomas spelled out the optimal cooperative process for deciding "what is good and necessarie for the common wealth."

When the king decided to challenge the sharing of power with Parliament, civil war was waged (1642 to 1646) and Parliament won. King Charles I was executed, and although his son soon returned to a restored throne, the monarchy would never be the same, for Parliament had become entrenched and democracy would steadily increase its hold over England. The practical transfer of power from monarchy to democracy went hand in hand with a reasoned inquiry into how the powers of government should be distributed. John Locke pioneered this analysis; he was a tutor at Oxford and a diplomat. He fled to Holland to avoid prosecution for sedition by James II. After the overthrow of James, Locke returned to England accompanying Princess Mary of Orange, who was to join her husband William on the throne in the Glorious Revolution of 1688, "glorious" because there was no civil war. The new constitutional monarchy fostered religious tolerance with increased freedom of thought, freedom of speech, and freedom of the press.

The European Enlightenment spread the new way of thinking. "Doubt and confusion eventually gave way to self-confidence, the belief that the unknown was merely the undiscovered, and the general assumption—unprecedented in the Christian era—that man was to a great extent the master of his own destiny" *(3)*. More specifically, the philosophers of the Enlightenment were working out how government operated, just as scientists of the Enlightenment were working out how the physical world operated. In 1748 Montesquieu wrote: "In every government there are three sorts of power: the legislative; the executive in respect to things dependent on the law of nations; and the executive in regard to matters that depend on the civil law. . . . the latter we shall call the judiciary power, and the other simply the executive power of the state" *(4)*.

The final blow to the British monarchy came in 1775 when the American colonists rebelled, and in so doing issued the first and probably the most eloquent statements of the new political ideas on how modern people should be governed in the Declaration of Independence and the Bill of Rights. Similar proclamations soon became the rallying cries of the French Revolution. The people's call for liberty, equality, and fraternity would ultimately wrest power from all the monarchies.

WHY DID MONARCHIES work so well, for so many years? Monarchies thrived where there was a population of uninformed and powerless subjects who had little option but to cooperate, in return for which they were protected from internal lawlessness and external attack. The most rational form of government for such a society was firm, centralized leadership—that is to say, monarchy. But the monarchies had trouble in dealing with the growth of commerce and knowledge. We can conclude, simply and rationally, that communities prosper when they choose the form of government that is most appropriate for their circumstances, but their circumstances change.

If we approach history hoping to find decreasing oppression and increasing freedom, we might be tempted to claim that progress has been achieved as the growth of civil liberty has led to fairer and more open societies. But the claim is misleading. Our political systems have been responsible for massive brutality over the course of the twentieth century; indeed, the record has been as bad as in any previous period. The evidence is more in keeping with a biological underpinning of history—the animal kingdom, including *Homo sapiens,* is continually adjusting to change, such as climatic trends and new social challenges. Democracy is proving to be the most versatile and successful political system for dealing with multicultural populations composed of literate people, many of whom have acquired power through commerce.

DEMOCRACY

The first well-documented democracy was in Athens some 2,500 years ago. The background is not entirely clear, but it seems that the nobility managed to gain power from the monarchs about 200 years earlier, and the high priority assigned to literacy created a well-informed section of the community extending beyond the aristocracy. Overseas trade flourished and this commerce brought money and therefore power to members of the newly formed "middle class." Because of an increased availability of metals, they were able to equip themselves as hoplites—heavily armed infantry. So a significant body of citizens gained power physically, financially and intellectually—a formula for social unrest. People began to make rational reappraisals of how they wanted to be ruled. In about 600 B.C., Solon was appointed to revise the constitution of Athens. As a poet, merchant, and magistrate, he was well equipped to reform the criteria for eligibility to hold government office, and instead of keeping influential positions for the aristocracy, he opened them up to the middle class. The monarchy and the aristocracy were forced to share their power and new words were coined such as *isonomia* (government based upon equality) and *demokratia* (power in the hands of the people). Public officials were chosen by election, or by the ultimate symbol of equality—the drawing of lots. The changes were far-reaching, but slaves remained slaves.

For about a century democracy flourished, then external forces intervened. Sparta crushed Athens in the Peloponnesian War and authoritarian government returned. When Rome became the dominant world power, democracy took a new meaning—government by an elite, achieved by placing strict limitations on who could vote and for whom. "Democracy" came to signify nothing more than a self-governing republic, and its original essence, the role of "the people," was lost. The social experiment conducted in Athens 2,500 years ago showed that democracy could work, but in the setting where it was born it proved too fragile for its time.

Subsequently a few other small countries tried decentralized government without a monarch, notably Iceland, Switzerland, and parts of northern Spain. But for the vast majority of humankind, kings ruled the world until the Enlightenment. Then frustration boiled over and philosophers assembled theories to back up actions. John Locke saw the human capacity for reason as justification for the diffusion of power to solve the problems of government. "The freedom then of man, and liberty of acting according to his own will, is grounded on his having reason which is able to instruct him in that law he is to govern himself by" (5). For Locke, reason leads to democracy: "Whosoever, therefore, out of a state of nature unite into a community must be understood to give up all the power necessary to the ends for which they unite into society to the majority of the community" *(5)*. And again: "There can be but one supreme power which is the legislative, to which all the rest are and must be subordinate, yet the legislative being only a fiduciary power to act for certain ends, there remains still in the people a supreme power to remove or alter the legislative when they find the legislative act contrary to the trust reposed in them" *(5)*.

The task of replacing the monarchies was fraught with difficulty. In the words of Montesquieu:

> One would imagine that human nature should perpetually rise up against despotism. But, notwithstanding the love of liberty, so natural to mankind, notwithstanding their innate detestation of force and violence, most nations are subject to this very form of government. This is easily accounted for. To form a moderate government, it is necessary to combine the several powers; to regulate, temper and set them in motion; to give, as it were, ballast to one in order to enable it to counterpoise the other. This is a masterpiece of legislation, rarely produced by hazard, and seldom attained by prudence *(4)*.

Montesquieu's analysis is typical of the political thought arising out of the Enlightenment. But the buildup of rational pressure for increasing liberty was, predictably, met by rational opposition; reason was recruited to discredit reason. Edmund Burke's conservative treatise *Reflections on the Revolution in France (6)* (1791) defended the monarchy, the aristocracy, the clergy, and even the new merchant class. Burke's attack was not focused on the masses, but on the intelligentsia, who fashioned the ideological backbone of the French Revolution. Reason was the tool of the men with the new ideas, but it was equally a tool that could be turned against them by traditionalists. The success of the counterattack is illustrated by the fact that a century ago, only 20 percent of the population of Europe could vote.

So rational arguments were used by both sides as the new political order overtook the *ancien régime*. Whether the monarchy disappeared suddenly or slowly, the transfer of real political power was irreversible.

Democracy called for cooperation, so the needs of the individual could be balanced with those of the community and rational adjustments could be made in response to circumstances that were always changing. Monarchy was generally more conservative but still rational, satisfying itself with the task of holding the status quo because political change involved risk, as recognized by Machiavelli:

> It must be considered that there is nothing more difficult to carry out, nor more doubtful of success, nor more dangerous to handle, than to initiate a new order of things. For the reformer has enemies in all those who profit from the old order, and only lukewarm defenders in all those who would profit from the new order; this lukewarmness arises partly from the fear of adversaries, who have the laws in their favour; and partly from the incredulity of mankind, who do not truly believe in anything new until they have actual experience of it *(1)*.

Democracy is inherently more flexible than monarchy, because an elected government wants to be reelected and it knows that its reelection depends upon making well-chosen responses to new situations.

John Rawls argues that fairness is the raison d'être for democracy:

> The fundamental organizing idea of justice as fairness, within which the other basic ideas are systematically connected, is that of a society as a fair system of cooperation over time, from one generation to the next. . . . In their political thought, and in the dis-

cussion of political questions, citizens do not view the social order as a fixed natural order, or as an institutional hierarchy justified by religious or aristocratic values* *(7)*.

Of course democracy, like any other form of government, is subject to intrigue and corruption, but the backstabbing tends to be metaphorical, rather than the literally murderous conflicts that occur with monarchies and dictatorships. The distinctive drawback of democracy is inertia; decisive action is difficult when powerful sections of the community are irreconcilably pitted against each other. At times of economic or military crisis, this inertia can be catastrophic—the checks and balances of democracy become a burden that hampers resolute decisions, and promotes stagnation in circumstances that demand action. This situation sets the stage for dictatorship to seize the reins of government.

DICTATORSHIP

Because democracies cannot escape economic and military crises, dictatorships have been frequent. The list, for the last sixty years, includes Stalin, Mussolini, Hitler, Franco, Perón, Tito, Mao, Castro, Nasser, Sukarno, Nkrumah, Amin, Marcos, Pol Pot, Noriega, Suharto, and Saddam Hussein. In theory, government by dictatorship is appropriate in difficult times, for when a country is in turmoil the granting of absolute power to a popular leader is a rational approach to the challenge of taking controversial but necessary actions. If the leader is wise and fair, the outcome should be good, and many of the great ancient Greek philosophers favored the concept of enlightened tyranny. Sadly, we have witnessed one disaster after an-

---

* Rawls goes on to explain cooperation in a way that somehow resonates with the concepts of reciprocal altruism. He seems to hint at the biological roots of successful social interaction: "Cooperation involves the idea of fair terms of cooperation: these are terms that each participant may reasonably accept, provided that everyone else likewise accepts them. Fair terms of cooperation specify an idea of reciprocity. . . ." *(7)*.

other at the hands of dictators, because aspirants for absolute power seem to select themselves as individuals who are neither wise nor fair; at best they have been corrupt egotists, at worst ideological fanatics. Dictators make use of special social tools—propaganda and secret police—to manipulate the population, but whatever may be said about the character and consequences of dictators, they have usually used reason to apply their policies.

Failure to use reason in the day-to-day operation of governments, including dictatorships, soon leads to insoluble problems and inevitable collapse. Perhaps the best recent example of irrationality is the administration of Idi Amin Dada of Uganda. History records a series of capricious and incoherent decisions culminating in economic disaster and bewildering brutality. In September 1972 a London newspaper summarized Amin's performance in the memorable headline: "He's Nuts." In a top secret memorandum one of his own cabinet ministers felt he had to put the facts on paper:

Amin finds it well nigh impossible to sit in an office for a day. He cannot concentrate on any serious topic for half a morning. He does not read. He cannot write. The sum total of all these disabilities makes it impossible for him either to sit in the regular Cabinet, to follow the Cabinet minutes, or to comprehend the briefs written to him by his ministers. In short, he is out of touch with the daily running of the country, not because he likes it but because of illiteracy. He rarely attends Cabinet and even then it is only when he is giving direction about problems concerning defense or security of the country or when he is sacking more civil servants *(8)*.

Equally revealing is a vignette of Amin's dictatorship in action:

A series of killings took place in December 1973. A woman who had gone to Mulago morgue to identify the body of a cousin shot by soldiers on 24 December found the bodies of about 50 people, all of whom had been shot. When she asked the morgue attendant what had occurred, she was told she was lucky she had come on a quiet day. A few days earlier a Baganda lawyer was briefed to defend a man in custody charged with embezzlement. He went to court with his client and applied for bail which was granted. Then as they walked out of the court the man was shot dead by men of the Public Safety Unit and the lawyer beaten unconscious. He was detained and flogged for four days before escaping to Kenya. The magistrate who had granted bail only escaped by jumping from a court room window *(8)*.

In contrast, most dictators have operated more consistently; their contributions to history have been rationally planned and rationally implemented, with the rational use of emotional incentives when necessary. The skillful blending of reason and emotion is the art of politics, and to see how strong dictators have operated we shall look at the most extreme and disastrous experiment in totalitarianism, the Third Reich of Germany.

Because his impact on the world was so enormous, Adolf Hitler has probably been scrutinized more than any other political figure of the twentieth century. We can glean little from his rather ordinary family background and childhood, but it is useful to recognize that Europe was seething with political activity when Hitler was young. At his secondary school in Linz, Austria, "The pupils flaunted in their buttonholes the blue cornflower popular among German racist groups. They gave preference to the colors of the German unity movement, black-red-gold; they greeted one another with the Germanic *'Heil!'* and sang the tune of the Hapsburg imperial anthem with the text of 'Deutschland über Alles' " *(9)*.

Hitler was a product of this nationalist and racist environment; he did not create it. As a corporal in the First World War he enjoyed the fighting, in which he displayed considerable courage. After the war he became actively involved in extremist politics, assembling his views into an implacable composite of ideologies that would later be expressed in *Mein Kampf* and the manifesto of the Nazi party. In 1920 he made his first speech to a mass meeting and discovered his awesome power as an orator, and he recognized how much the German people yearned for a strong leader. For the next thirteen years he harnessed reason and emotion to his task of becoming Reich chancellor. If he made a mistake, such as the failure of his revolutionary *Putsch* in Munich in 1923, he learned from the experience and did not repeat it. He applied the old technique of political marches through the streets, and he pioneered the new communication media of radio and films. He also pioneered the use of aircraft to make dramatic political tours across the country.

By means of ritual and torchlight processions he was able to raise the emotional fervor of rallies to the level of religious experiences. He was a master of rhetoric who had no difficulty in whipping his audiences into an ecstasy of pride in "belonging"—and in filling them to the brim with a burning hatred for those who he decided did not "belong"—the Jews. He carried both the carrot of glory for those who would follow him and the stick of violence for those who would not.

Hitler planned and executed his ascent to power through the rational dissemination of emotionally charged agitation. He used reason to orchestrate the techniques of propaganda in a climate of economic crisis. He became Reich chancellor in January 1933 without breaking any law or breaching any constitutional protocol. He then dissolved all rival political parties in July 1933, and the following year he became Reich president and supreme commander of the armed forces.

Hitler's foreign policy was equally rational and successful. He extended German power by occupying the Rhineland, annexing Austria, and dismembering Czechoslovakia. In 1939 he declared, with some justification: "I overcame chaos in Germany, restored order, enormously raised production in all fields of our national economy.

... I succeeded in resettling in useful production those seven million unemployed who so touched all our hearts. ... I have not only reunited politically the German nation but also rearmed it militarily" *(10)*. In a penetrating analysis of Hitler, Sebastian Haffner draws a picture of a dictator who was persistent and entirely ruthless. His achievements could not be denied, and they fueled his popularity:

> When, in the twenties, Hitler had at his disposal nothing but his demagogy, his hypnotic oratory, his intoxicating and illusionist skills as a producer of mass spectacles he hardly ever gained more than 5 per cent of all Germans as his followers; in the Reichstag elections of 1928 it was 2.5 per cent. The next 40 per cent were driven into his arms by the economic plight of 1930–33 and by the total helpless failure of all other governments and parties in the face of that plight. The remaining, decisive, 50 per cent, however, he gained after 1933 mainly through his achievements *(10)*.

In 1939 Hitler jeopardized Germany's no longer disputed hegemony in Europe by converting it to the military conquest and occupation of Europe, "a step comparable to the deliberate rape of a woman perfectly willing to surrender"* *(10)*.

How did the German people come to accept Hitler as their dictator? When the Weimar Republic was unable to control hyperinflation, Hitler offered hope, and he restored pride to a community that was still tormented by the loss of the war and the humiliating Treaty of Versailles. He also provided an explanation for all the failures of the past in terms of a stab in the national back by the Jews, in a climate where the culture would accept this because of past anti-Semitism embelished by contemporary Nazi propaganda. But Hitler was not brought to power through the misguided will of the uncrit-

---

* Hitler had to declare war because he wanted more than Europe; he wanted the vast territory extending from Warsaw to Moscow.

ical and uninformed masses; his most enthusiastic supporters included a great number of the intelligentsia—writers, physicians, lawyers, and those with the highest training in reason, the university professors. The most famous German philosopher of the twentieth century, Martin Heidegger, was a Nazi. People followed Hitler because of the circumstances in which they found themselves, and the prevailing cultural attitudes. But while Germany, between the wars, provided a fertile soil for Hitler to grow, he had enough support from neighboring countries to be regarded, historically, as a European, rather than just a German phenomenon.

FROM THE START of his climb to power, Hitler clearly intended to resume the world war that had started in 1914 and stopped, for him prematurely, in 1918. The only question was when. He chose 1939 because he wanted to have all his energy and wits about him for the armed struggle that he saw as the high point and ultimate purpose of his life. He may also have started to detect a decline in his health due to the insidious onset of Parkinson's disease, as we shall discuss later. Hitler was so set on war in 1939 that "he told his military adjutant, Major General Engel, in August he was consumed with anxiety lest some 'blöder Gefühlsakrobat mit windelweichen Vorschlägen' (stupid acrobat of the emotions with wishy-washy proposals) would thwart him at the last moment" *(11)*. Once war broke out, Hitler's personal direction of the Battle of France was rational and flawless; his achievements as a military commander only started to decline after his early successes in Russia. Then the military situation deteriorated rapidly. By the end of the Battle of Stalingrad Hitler must have known that he could not win the war. Why, then, did he not attempt to negotiate an end to the fighting? After all, he had previously signed (and broken) the Soviet-German nonaggression pact, which was forged in highly unlikely circumstances, so why not make the rational decision to save what he could for his German nation-state while he still had some bargaining points? Nothing could be worse than slow, remorseless military defeat. Surely he must have thought of how history would judge him if he made no attempt to avert the total destruction of his country, not to mention his place in history if all the horrors of the extermination camps were laid bare, as they were bound to be if the Allied military onslaught was allowed to continue.

These are intriguing questions that direct us to the core of Hitler's personality. He had achieved so much because he was resolute, but resolute is an understatement: he was totally single-minded when it came to his intense beliefs. His ruthlessness can be seen in the routine directives he issued, for example the "Führer's Decree" that gave the armed forces immunity for crimes against civilians. In the same vein, he issued an order that commissars of the Red Army, "when captured in battle or resistance are on principle to be disposed of by gunshot immediately." Given this evidence of his extraordinary intransigence, the usual explanations of why he continued the war seem superficial. As his armies retreated on all fronts he knew that he and Germany were doomed and defeat was inevitable. He was too astute to have harbored any real expectation of miracles from German secret weapons such as rockets and jet aircraft, or a naive hope for a break up of the alliance between Great Britain, the United States, and Russia. It is true that the Allied ultimatum for unconditional surrender posed a problem. But whether he surrendered or fought on, his personal future was clear—sooner or later he would have to take his own life. We must look further, at his most deeply held convictions, to find why he did not save what he could of his fatherland in the later stages of the war.

Hitler's strongest convictions, combined with his unshakable determination and ruthlessness, afford a plausible explanation of what happened. What were his strongest convictions? When he first came to Vienna as a young man, there were numerous Austrian and German nationalist groups he could join, yet his first choice, over and above all the nationalist organizations, was The League of Anti-

Semites. He embraced anti-Semitism without reservation and his anti-Semitism was extreme—in fact, it was murderous. He remained so consumed by it that on the last day of his life, he ended his dictated political testament exhorting the German people to meet their ultimate responsibility: "Above all I charge the leaders of the nation and those under them to scrupulous observance of the laws of race and to merciless opposition to the universal poisoner of all peoples, international Jewry."*

Many of Hitler's attitudes are not accessible to analysis, but his anti-Semitism can be discredited by reason. Let us assume that Hitler wanted to further the interests of Germany by increasing its prosperity, its prestige, its influence, and the extent of its territory. If this is a valid starting point, his attack on the Jewish community flew in the face of reason. Hitler argued that the major needs of Germany, when he took power, required a supreme national effort, and that some of the observed attributes of the Jews—their independent traditions and clannishness—conflicted with national unity. But it was clear that other observed attributes of the Jews, such as their skills in science, medicine, and commerce, were highly desirable. Given the pros and cons of accepting the Jewish community as German citizens, Hitler's initial anti-Semitic policies were distinctly irrational, for the benefits brought by the Jews unquestionably outweighed the drawbacks. Of course, the escalation of virulent hatred to the extent required for genocide magnified the irrationality by breaching a universal and rational ethical edict—against murder.

From the picture we have of the intensity of Hitler's anti-Semitism, Haffner concludes that when the war started, the first priority was "Victory for Germany at any cost," but when it became certain that this was impossible, Hitler's next priority became "Death

---

* This translation comes from Lucy S. Dawidowicz's book *The War against the Jews 1933–1945* (New York: Bantam, 1976). This book, like many others, searches for some personal incident to account for Hitler's hatred, but nothing convincing stands out.

for the Jews at any cost" *(10)*. And any cost included prolonging the war, with the consequent loss of hundreds of thousands of his own people, together with the physical devastation of his country. This analysis of Hitler's motivation is in keeping with his order to kill six million Jews who could otherwise have been used for forced labor, and his decision to continue to operate the death camps and their transport systems when men and rolling stock were urgently needed at the front. In the words of Haffner: "To Hitler, during the last three and a half years of war, the war had become a kind of race which he was still hoping to win. Who would reach his goal sooner, Hitler with his extermination of the Jews or the Allies with their military over-throw of Germany?" *(10)*.

While Hitler's anti-Semitism was irrational, he used highly rational methods to pursue his goal. It is both disturbing and instructive to see how effectively a dictator could engage reason in a fanatic enterprise that operated in competition with the war effort and in conflict with established morality. The Holocaust was not the first attempt at genocide, but its scale was unprecedented. How was it possible for Hitler to involve so many Germans in the organized slaughter of millions of innocent civilians? The practical difficulties of the task meant that mere manipulation of popular emotions was not enough; reason was required for the "Final Solution"—Hitler needed highly skilled bureaucratic planning and highly efficient technical implementation.

First, there was the challenge of finding the Jews. Responsibility for solving this problem was given to the Gestapo, an institution that represented "the most sinister and horrific aspects of Hitler's dicta-torship, including arbitrary arrest and detention, endless interroga-tions, forced confessions under torture in police basements, and willful misuse of police authority" *(12)*. In retrospect, many think of the Gestapo as a massive infiltrating agency made up of extreme Nazis who spied on a cowering and submissive population. From ex-tensive studies of the records, Robert Gellately paints a very different picture of the secret police: "Most were not initially members of the Nazi party, but were trained policemen, and many were carry-overs from the Weimar days. It was a much smaller force than is sometimes suggested, a kind of high level group of 'experts' "; it had few secret informants, and it relied heavily upon reports provided by the general public. In general, the problem for the Gestapo was not how to get the public to denounce Jews and dissidents, but how to sift through the many false charges.

The motives for offering information to the authorities ranged across the spectrum from base, selfish, personal, to lofty and "idealistic." . . . Successful enforcement of Nazi racial policies depended on the actions of enough citizens, operating out of an endless variety of motives, who contributed to the isolation of the Jews by offering information to the Gestapo or other authorities of Party or state. The other side of the coin, the astonishingly rare remarks about the Nazi terror, and especially the few occasions on which people broke their silence on the persecution of the Jews in the country, provide additional, silent testimony as to the effectiveness of the enforcement of racial policy *(12)*.

Once Jews had been arrested, they had to be killed. In his book *Hitler's Willing Executioners: Ordinary Germans and the Holocaust,* Daniel Goldhagen writes: "It has been generally believed by scholars (at least until very recently) and non-scholars alike that the perpetrators were primarily, overwhelmingly SS men, the most devoted and brutal Nazis. It has been an unquestioned truism (again until recently) that had a German refused to kill Jews, then he himself would have been killed, sent to a concentration camp, or severely punished" *(13)*. Yet Goldhagen points out, "Life within the camp system demonstrated how radically ordinary Germans would implement the racist, destructive set of beliefs and values that was the country's formal and informal public ideology."* The technical problems of killing and disposing of the bodies of millions of people were confronted methodically. Scientists were encouraged to conduct experiments on live human biological material before it went to waste. After an unsatisfactory experience with carbon monoxide in early "euthanasia" programs, hydrogen cyanide was found to be the most cost-effective agent for killing large numbers of people; a safe and reliable canister system was developed for delivering it into carefully constructed chambers. Engineers met the challenge of designing and constructing equipment such as incinerators that would fuel themselves from the melting fat of human bodies; everything was worked out rationally for the job in hand.

The key to Hitler's success in pursuing his racial policies is the unpalatable but undeniable fact that by 1940 he was a very popular leader. Because of his popularity he was able to take preexisting cultural attitudes and intensify them unremittingly until they reached the extreme level required for genocide. Reason gave him the propaganda machine to promote his goals, but the goals themselves had nothing to do with reason. Hitler's ideology was built upon the mythology of "Aryan" racial superiority. Naturally, it had a strong appeal to many Germans—the call to "cleanse the Fatherland" proved to be such a powerful slogan that everything else could be sacrificed for it.

The Nazi racial policies were built upon the argument that the German *Volk* are the master race. To examine this claim, we must give priority to observation over theory—that is to say, we must use reason in the form of Galileo's knife. The notion of a master race is open to verification or falsification by observation. While the Ger-

---

* Goldhagen's book has raised a controversy among historians. He argues that the central cause of the Holocaust was a peculiarly intense and enduring German hatred for Jews. This seems too simple, for other European countries had a history of deep anti-Semitism. Recently, some of Goldhagen's factual assertions have also been questioned. See N. G. Finkelstein and R. B. Birn, *A Nation on Trial* (New York: Henry Holt, 1998).

mans have achieved much over the last 500 years, their success in philosophy, the arts, the sciences, technology, sport, warfare, and other
human endeavors has been no greater than, for example, the French,
the Americans, or the British. Reason forces the concept of a German
master race into the same category of untruth as the claim that the
earth is flat.

Hitler's attitudes on race, and the acceptance of these attitudes by
his government, are a poignant illustration of the primacy of feelings
over facts in setting political goals. Hitler was exceptionally charis-
matic, ruthless, and vicious, but his cultural makeup was not especially unusual; his views reflected feelings of arrogant superiority
and intolerance that were common among the intellectual community of his time. John Carey argues that Hitler's general opinions
were shared, to at least some extent, by such members of the international intelligentsia as Wyndham Lewis, D. H. Lawrence, H. G.
Wells, T. S. Eliot, and Ezra Pound. As a macabre example, Carey cites
D. H. Lawrence writing to Blanche Jennings in 1908: "If I had my way,
I would build a lethal chamber as big as the Crystal Palace, with a
military band playing softly, and a Cinematograph working brightly;
then I'd go out into the back streets and main streets and bring them
in, all the sick, the halt, and the maimed; I would lead them gently,
and they would smile me a weary thanks; and the band would softly
bubble out the 'Hallelujah Chorus.' " Carey quotes him again: "Three
cheers for the inventors of poison gas," and, in 1921: "I don't believe
in either liberty or democracy. . . . I believe in actual, sacred, inspired
authority" (14). The Germans were certainly not alone in their attitudes.

While dictators use reason extensively to further their aims, they
have special expertise in arousing emotions, and a natural gift for
large-scale crimes against humanity because of their affinity for politically extreme ideologies and their ability to pursue these ideologies without accountability. Charisma and ruthlessness are useful
traits for capturing popularity and power but they are also the traits

of psychopathy. In 1933, a well-known German psychiatrist, Ernst Kretchmer, wrote in a letter: "It's a funny thing about psychopaths. In normal times we render expert opinions on them. In times of political unrest they rule us"* *(15)*.

---

* Kretchmer is best known for his studies on the relationship between mental disease, such as schizophrenia, and body build.

8

REASON FOR RELIGION

*Nobody can deny but religion is a comfort to the distressed, a
cordial to the sick, and sometimes a restraint on the wicked;
therefore whoever would argue or laugh it out of the world
without giving some equivalent for it ought to be treated as a
common enemy.*

—MARY WORTLEY MONTAGU*

WHY ARE WE here? Why do we suffer? Why do we die? What
happens after death? These questions all have a similar ring because
none can be answered by the usual method of looking at how the
world works. They defy the normal methods of verification, yet they
seem to *demand* answers. It is not simply that we *want* answers, we
seem in some way to *need* answers. Religion provides them.

In the words of Melford Spiro:

Religion persists because it has functions—it does, or is believed
to, satisfy desires; but religion persists because it has causes—it is
caused by the expectation of satisfying these desires. Both are nec-

---

* Letter to her daughter, Lady Bute, June 23, 1754. In *The Complete Letters of Lady Mary
Wortley Montagu*, ed. R. Halsband. Oxford: Clarendon Press, 1965–1967.

essary, neither is sufficient, together they are necessary and sufficient. The causes of religion are to be found in the desires by which it is motivated, and its functions consist in the satisfaction of those desires which constitute its motivation *(1)*.

Religion is of such importance to people that it has become a dominant cultural institution. It has assumed responsibility for retaining our most profound wisdom, establishing our morality, and bonding individuals into a communal unity. It also provides happiness to many and ecstasy to a few. The building blocks of religion are simple assertions and a belief that we can influence the future, through prayer. The articles of faith come from intuition, inspiration, and revelation; they are then codified into authoritative statements that can be passed from one generation to the next. Reason is used for the operation of religious institutions that possess priests, temples, and rituals. Religious scholars also use reason to bring religions up to date with developments in the mundane world.

Avicenna (980–1037) was the first great mind to come to grips with the challenge of searching for reason within the doctrines of a formally structured and highly influential religion. He was Persian, and by the age of ten years he had memorized the Koran; over the course of his life he was to write some two hundred works, the longest being *The Book of Healing* (translated into Latin in the twelfth century). Avicenna's philosophy was influenced by Aristotle, and like Socrates he claimed he could prove, through reason, the existence of an immortal soul. He earned his living as a physician, and his medical texts continued in use several centuries after his death.

The next great scholar to pursue reason in religion was another physician, Maimonides (1135–1204); he was born in Spain, but as a Jew he was forced to leave because of persecution by Islamic extremists. Maimonides spent most of his life in the old city of Cairo, and just like Avicenna, he depended on his medical practice for income.

His philosophical masterpiece was *The Guide of the Perplexed,* about which Paul Johnson writes:

> At every stage, in the code and the commentary, he was ratio-
> nalizing. But in addition, he wrote his *Guide of the Perplexed* to
> show that the Jewish faith was not just a set of arbitrary asser-
> tions imposed by divine command and rabbinical authority, but
> could be deduced and proved by reason too. . . . In his *Guide of
> the Perplexed,* he sets out his intensely rationalistic view of
> the *Torah:* 'The law as a whole aims at two things—the welfare of
> the soul and the welfare of the body.' The first consists in devel-
> oping the human intellect, the second in improving men's polit-
> ical relations with each other. The Law does this by setting down
> true opinions, which raise the intellect, and by producing norms
> to govern human behavior. The two interact. The more stable
> and peaceful we make our society, the more time and energy
> men have for improving their minds, so that in turn they
> have the intellectual capacity to effect further social improve-
> ments *(2).*

In a similar way, St. Thomas Aquinas (1225–1274) found reason in religion. In the words of Bertrand Russell, Aquinas

> maintained—and his view is still that of the Roman Catholic
> Church—that some of the fundamental truths of the Christian re-
> ligion could be proved by unaided reason, without the help of rev-
> elation. Among these was the existence of an omnipotent and
> benevolent Creator. From His omnipotence and benevolence it
> followed that He would not leave His creatures without knowl-
> edge of His decrees, to the extent that might be necessary for obey-
> ing His will. There must therefore be a Divine revelation, which,
> obviously, is contained in the Bible and the decisions of the
> Church. This point being established, the rest of what we need to

know can be inferred from the Scriptures and the pronouncements of ecumenical Councils. The whole argument proceeds de-
ductively *(3)*.

These claims of rationality echoed through the following centuries. Countless scholars devoted their lives to exhaustive analysis,
interpretation, and reconciliation of religious texts. After the Renaissance, reason was engaged more widely for religious purposes. In
1696 the English theologian William Whiston wrote a book entitled
*A New Theory of the Earth; wherein the Creation of the World in Six
Days, the Universal Deluge, and the General Conflagration, as laid
down in the Holy Scriptures, are shown to be perfectly agreeable to Reason and Philosophy.*

A notable nineteenth-century high-church Anglican, Henry
Mansel, continued the scholarly blending of reason and religion in
a series of lectures published as *The Limits of Religious Thought.*
Mansel's lectures received national press coverage. The *Times* reported: "Sunday after Sunday, during the whole series, in spite of the
natural craving for variety, and some almost tropical weather, there
flocked to St. Mary's a large and continually increasing crowd of
hearers, to listen to discourses on the Absolute and the Infinite,
which they confessedly could not comprehend."* Mansel presented
a modern argument for the infallibility of the Bible. The capacity of
the human mind is inherently limited and religious dogma tran-
scends these limits. Therefore dogma is inscrutable and no observation can be called upon as evidence against it. In essence, Mansel said
you either have faith or you do not, because the human mind does
not have the capacity to confirm or refute religious precepts. So no

---

* Mansel was appointed the first Wayneflete Professor of Moral and Metaphysical Philosophy at Oxford in 1855, and in 1868 he became Dean of St. Paul's Cathedral. He published *The Limits of Religious Thought* in 1868. His impact on nineteenth-century religion
and irreligion is recorded in Bernard Lightman's *The origins of Agnosticism* (Baltimore:
Johns Hopkins University Press, 1987).

attack on dogma can be meaningful or rational. The argument is persuasive: there are matters of great importance that we discuss and debate although they are inaccessible to rational confirmation or refutation. Mansel's conclusion was reached by reasoned argument, even if dogma, in general, must be accepted or rejected without recourse to reason.

The relationship between reason and religion continues to attract attention. *Faith and Reason* is the title of John Paul II's thirteenth encyclical (1998). The pope denounces "the fateful separation of faith and reason in modern times." He asks the church to "lead people to discover both their capacity to learn the truth and their yearning for the ultimate and definitive meaning of life."*

IN ADDITION TO having a role to play in the theory of religion, reason also contributes to the practice of religion. The broad brush strokes of religious dogma have to be followed by fine brushwork when it comes to dealing with detailed questions, and reason is required for this transition. Take, for example, the Hindu teaching that cows are sacred. The *Mahabharata* states: "All that kill, eat and permit the slaughter of cows rot in hell for as many years as there are hairs on the body of the cow so slain." If cows are holy, we must protect them. Reason is needed to help translate the idealized view of cows into detailed practical guidance on how to manage cows in our daily life. Should we feed them before ourselves? Can we drink their milk? Can we use their skins to make clothes and shoes?

Religion has to answer every question on how we should live, and reason brings consistency and coherence to the answers. By looking at the features shared by different religions we can gain some understanding of the rational framework into which they all fit. Religions promote faith in a supernatural power that can be approached

---

* To illustrate the speed with which old religions can adopt new technologies, the text of the thirteenth papal encyclical can be found at: www.vatican.va

through prayer, and they inspire states of enhanced mental experience described, in their most intense form, as ecstasy. Every religion has stories about events, people, and places in its early history. Religion generates standards of required behavior and disseminates them as an ethical system in the form of commandments. Often the starting point is dogma—a statement of the innate depravity of humanity. In Abraham Maslow's view, this premise "leads to some extra-human interpretation of goodness, saintliness, virtue, self-sacrifice, altruism, etc. If they can't be explained through human nature—and explained they must be—then they must be explained from outside of human nature. The worse man is, the poorer thing he is conceived to be, the more necessary becomes a god" (4).

More practical commandments spell out rational ethical guidelines for sustaining a culture in which people can work together by cooperating and respecting each other's needs. Reason plays an essential part in creating this moral code. Religion also gains strength by bringing people together and nurturing a sense of belonging. As Maslow declares: "basic human need can be fulfilled *only* by and through other human beings, i.e., society. The need for community (belongingness, contact, groupiness) is itself a basic need. Loneliness, isolation, ostracism, rejection by the group—these are not only painful but pathogenic as well" (4). There is a positive feedback loop. Membership in the religious group entails obligations, in return for which the group, or more precisely the group's religion, will safeguard the interests of its members. Ultimately, this relationship creates powerful commitments—people become ready to sacrifice their self-interest, if necessary their lives, for their religions.

Most religions have congregations led by priests who practice divinely ordained rites in temples or shrines. The priests work within institutions that are just as rational as their secular counterparts. The responsibilities of many religions are extensive: they operate highly organized bureaucracies, and they build places of worship of extreme beauty, often famous for their art and technology. They often run schools, hospitals, and charities. Religions have made important contributions to the rational development of agriculture, and their financial expertise has been second to none. Most would acknowledge a central role for reason in their achievements.

ANCIENT RELIGIONS*

Knowledge of the earliest religions comes from archeological findings. Death has such an important place in all faiths that if we had to choose a single pivotal facet of religious practice to study it would probably be funereal rites, and this is where archeology happens to yield very good evidence.

The Cult of Skulls. The first detectable religion, going back 300,000 years, has been termed the *cult of skulls.* Skulls have been found, repeatedly, separated from other skeletal remains. The top vertebrae are usually missing from the base of the skulls, indicating that heads were removed after death—decapitation during life leaves the top vertebrae firmly attached to the skull. Bodies must have been buried initially and allowed to decompose, then the skulls could be removed without the top vertebrae. The way the skulls were finally arranged suggests that they were endowed with supernatural significance. A Neanderthal skull discovered near Monte Circeo, Italy, was deposited in the inner chamber of a grotto used for the organized storage of bones 100,000 to 70,000 years ago. This skull was placed on top of small fractured bones of an ox and a deer. Around the walls of the chamber were piles of bones from horses, hyenas, deer, lions, and elephants. In Bavaria, a Mesolithic group of twenty-seven human skulls was found embedded in red ochre, perhaps a symbol of blood and

---

* We will be looking at how religion changed over history, but this chronological approach is not intended to suggest that ancient religions were inferior. Old religions were best suited to old cultures. Religions changed in response to the new needs of larger communities, and they expanded through conquest, colonialization, and conversion.

life. The skulls were oriented so that they faced west, and a similar arrangement of six skulls was discovered a few yards away. In these Bavarian cases, decapitation had been performed on fresh corpses, so perhaps the heads were sacrificial. At some sites skulls were adorned with shells and there is evidence that fires were lit nearby.

We can only imagine the rites that led to the creation of these archeological artifacts, but the events must surely have been religious in nature. How can we interpret the cult of skulls in our search for the role of reason in early religion? Its extensive scale tells us something; reason must have been involved for a practice to be sustained over such a wide geographical area for such a long period of time. It is also tempting to speculate that reason may have contributed to the focus of religious attention on the head. Why was the skull deemed sacred? Our ancestors may have decided that the head was all-important and somehow held the very essence of being human. The head contains most of the sensory apparatus—for vision, hearing, taste, and smell; the significance of the sense organs is self-evident. The head is also the prime organ of communication, through facial expression and speech—perhaps this is why it was given privileged status. S. G. F. Brandon suggests: "The face was esteemed as the mirror of the individual self, which in turn would imply that individual characterization was realized and appreciated, and. . . . the head was regarded as the seat and center of the individual's life, both physical and psychical" (5).

The ancient Egyptians mummified the body as a whole, but some of their religious writings drew special attention to the head—life entered the body through the left ear and departed through the right one. Until quite recently Hindu cremation ceremonies gave similar priority to the head. After the funeral pyre was ignited and the body started to burn, the soul was thought to be trapped in the head. Normally the intense heat boiled the brain and the rise in intracranial pressure shattered the skull, setting the soul free. If the skull did not explode in this way, it would be split open with a cudgel. Head hunting persisted in Borneo until the Second World War. In 1871 E. B. Tylor described how the Dayaks who lived there believed that

> every human head they could procure would serve them in the next world, where, indeed, a man's rank would be according to his number of heads on this [world]. They would continue the mourning for a dead man till a head was brought in, to provide him with a slave to accompany him to the habitation of souls; a father who lost his child would go out and kill the first man he met, as a funeral ceremony; a young man might not marry till he had procured a head, and some tribes would bury with a dead man the first head he had taken, together with spears, cloth, rice, and betel(6).

The Dayaks said that their preoccupation with heads was equivalent to the white men's preoccupation with books—there was just a difference in culture. The funereal significance of heads was not confined to Borneo; Tylor recorded that heads were taken as funereal offerings in northeastern India. He also speculated that the custom of taking scalps, among American Indians, might have its origin in the cult of skulls.

Interment and Cremation. With the passage of time, the ways of dealing with the dead changed; emphasis moved from the head to the whole body, and fertility became an important religious issue. In France the complete skeleton of a Neanderthal was found resting on a pillow of flint flakes, a hand axe, and scraper by the left arm. Split and charred bones of a wild ox were also present in the grave, presumably the remains of a funeral feast. The trend toward more elaborate interment continued. At another grave site in the Dordogne, cowrie shells lay in pairs at strategic locations around the body—on the forehead, the upper arms, the thighs, the knees, and the feet. The cowrie, considered by many to represent the vulva at this time, was also a probable symbol of fertility. This theme continued to a period, some 24,000 years ago, when small ivory and stone female figurines came into use, with large pendulous breasts, broad hips, and round buttocks. Some appear to be pregnant, and some are adopting a squatting posture suggestive of childbirth. These figurines may signify the existence of cult worship for a great mother goddess who reemerged as Ishtar in Babylonia, Isis in Egypt, and Artemis in Greece. Phallic symbols reinforced the ancient preoccupation with fertility.

Then a new wave of religious thought evolved along the Nile valley, focusing on the afterlife; ultimately the prospect of life after death would dominate the minds of the Egyptians. At Badari, small scattered groups of graves contained elaborately adorned corpses:

> One man had a girdle of over 5,000 blue glazed steatite beads, and in the hair of a baby were ostrich-tips, while near another infant were a bowl and an ivory spoon. Among the graves were cooking-pots and food bowls, either empty or containing grain and meat, suggesting that meals were probably a mortuary ritual. Near the hands were slate palettes and pebbles for grinding malachite as eye-paint. The corpse was . . . surrounded with beads, perforated shells, ostrich-shell discs, ivory hair-combs and female figurines, as well as with pottery (7).

With the establishment of the First Dynasty in Egypt, 5,000 years ago, the simple pit graves continued for ordinary people, but the royal tomb became the overwhelming national cult. The permanent preservation of mortal remains was an obsession. Resin, cedar oil, olive oil, honey, incense, wax, and sodium carbonate were all used in the embalming process. The brain was extracted through the nostrils, perhaps because techniques were inadequate for its preservation. The internal organs were removed, except for those that could be mummified—the heart and sometimes the kidneys. Certain animals were accorded special religious significance in Egypt, and creatures such as cats, dogs, falcons, bulls, and crocodiles were mummified.* Reason must have played a part in developing the highly successful technology of mummification. Reason would have been equally essential for the architecture, engineering, and construction of immense, elaborate tombs such as the Great Pyramid.

Burial customs for the general population changed again as Neolithic and bronze age cultures spread into Europe 5,000 to 3,500 years ago. Hundreds of bodies were placed in caves, and communal burials took place in barrows. Toward the end of the bronze age buried ships took Viking warriors on their long voyage from death to immortality.

In Asia, bodies were often buried twice; after the first burial they were dug up and the bones were cleaned and rearranged. Some graves contained the scapulae of animals and plastrons of turtles that had been scorched to produce cracked "oracle bones"; from the patterns of cracks the cognoscenti could read the future. Jade goblets were often buried with notable persons but sometimes inanimate objects were not enough. Early Chinese emperors were buried with human and animal sacrifices—the archaeological evidence suggests that servants were expected to wait upon their masters for a long time.

As the bronze age evolved into the iron age, cremation became prevalent. Ashes were deposited in urns and then buried in large cemeteries. Early urns were simple; they often had holes in their sides, perhaps to allow the "spirit" to escape. Later urns were more elaborate—some were shaped like vases with side spouts, while others were shaped like animals. The spread of cremation coincided with increasing faith in a celestial afterlife and the building of sky temples like Stonehenge.

---

* In the late nineteenth century mummies of animals were exported to England and pulverized for use as garden fertilizer.

So the focus of ancient religions moved from the head, to the body, to fertility, to an afterlife. For each new cosmology, the dogma of religion was assembled into a comprehensible form that fitted the needs of the culture. Reason was necessary for the operation of ancient religions—for the organization of priests and the interpretation of dogma. Rational methods must have been used for the technology of religion, including embalming, the manufacture of burial urns, the construction of tombs, and the alignment of sky temples in geometric relation to celestial landmarks.

EXISTING RELIGIONS

Primal Religions. The simplest classification of faiths is into the *primal religions,* which are generally very old, with no written traditions, and the *great religions,* which are relatively new, with extensive writings compiled into authoritative texts. The great religions have larger followings than the primal religions, but there is no evidence that they approach closer to ultimate truths.

The primal religions exist among the Australian Aborigines, the Kalahari Bushmen, the American Indians, and virtually every other community that managed to delay domination by one of the giant literate cultures. The primal religions are all-pervading; narratives are mixed with practical information such as how to hunt and fish, where to find plant foods, and how to manage sickness. Animals are often endowed with human qualities and vice versa—this totemism can be extended to plants and even to inanimate objects. Huston Smith describes what happened when Oren Lyons, the first Onondagan to enter college, returned to the reservation. His uncle took him fishing in a boat and cross-examined him:

Once he had his nephew in the middle of the lake where he wanted him, he began to interrogate him. "Well, Oren," he said, "you've been to college; you must be pretty smart now from all they've been teaching you. Let me ask you a question. Who are you?" Taken aback by the question, Oren fumbled for an answer. "What do you mean who am I? Why, I'm your nephew of course." His uncle rejected his answer and repeated his question. Successively, the nephew ventured that he was Oren Lyons, an Onondagan, a human being, a man, a young man, all to no avail. When his uncle reduced him to silence and he asked to be informed as to who he was, his uncle said, "Do you see that bluff over there? Oren, you *are* that bluff. And that giant pine on the other shore? Oren, you are that pine. And this water that supports our boat? You are this water" *(8)*.

A Navajo chant reinforces the same points:

The mountains, I become part of it . . .
The herbs, the fir tree, I become part of it.
The morning mists, the clouds, the gathering waters.
I become part of it.
The wilderness, the dew drops, the pollen . . .
I become part of it.

We can also go back in time to find hauntingly similar beliefs expressed, over a thousand years ago, in the poems of Taliesin, a Welsh bard of the late sixth to early seventh century. In "Can y Meirch" (Song of the Horses) he wrote:

Of the nature of animals I am aware;
A goat I was in the elder wood,
And a wild sow of the forest;
On a mountain peak as a buck I stood.

And I have been the self same mountain,
And the rain which on it fell,
To the stream which downward raging
Vanished at last in Ocean's swell.

As a brindled cat I climbed the tree-tops,
And far above as a crane I soared
Far-sighted then; but in the forest
As a panting predator I roared . . .*

The Australian Aboriginal people share these views on the opposite side of the world; their culture has been isolated for some 50,000
years, since the glacial period temporarily dropped the sea level and
allowed colonization by inhabitants of the Indonesian lowlands. "All
things on earth we see as part human. We see all things natural as
part of us."† The identity is so close that:

Tree and grass same thing.
They grow with your body,
with your feeling.
If you feel sore. . . .
headache, sore body,
that mean somebody killing tree or grass.
You feel because your body in that tree or earth.
Nobody can tell you,
you got to feel it yourself.

Tree might be sick. . . .
you feel it.
You might feel it for two or three years.
You get weak. . . .
little bit, little bit. . . .
because tree going bit by bit. . . .
dying.

---

* This translation was prepared by Dr. Hugh McLennan, University of British Columbia,
who kindly gave permission for its publication here.
† These two quotations come from *Kakuda Man* by Bill Neidjie, Stephen Davis, and
Allan Fox (New South Wales: Allan Fox and Associates, 1985). The first passage is attributed to Silas Roberts, and the second to Bill Neidjie.

For the primal religions, all of nature possesses a significance that can, in a sense, be termed sacred. For example, hunting requires a high level of physical and mental ability, but these attributes, by themselves, do not suffice. The Sioux shaman Black Elk, describes how a hunter

> does not set out simply to assuage his tribe's hunger. He launches on a complex of meditative acts, all of which—whether preparatory prayer and purification, pursuit of the quarry, or the sacramental manner by which the animal is slain and subsequently treated—are imbued with sanctity. A reporter who lived with Black Elk for two years recounts the latter's assertion that hunting *is*—Black Elk did not say represents, the reporter emphasized—life's quest for ultimate truth; the quest requires preparatory prayer and sacrificial purification *(8)*.

The final contact with the quarry is the ultimate realization of "truth."

The primal religions dwell on biological functions, such as eating, drinking, sexual intercourse, childbirth, fighting, killing, and dying. Even breaking wind can be immortalized in religious narratives. Lévi-Strauss tells the traditional story of a disobedient Bororo Indian boy and his grandmother.

> Irritated by his behavior, his grandmother came every night while he was asleep and, crouching above her grandson's face, poisoned him by emissions of intestinal gas. The boy heard the noise and smelled the stench, but he did not understand where it was coming from. Having become sick, emaciated, and suspicious, he feigned sleep and finally discovered the old woman's trick. He killed her with a sharp pointed arrow which he plunged so deeply into her anus that the intestines spurted out *(9)*.

The stories survive from generation to generation without the written word because a highly rational process of indoctrination—by endless repetition, ceremonial chanting, dancing, and reenactment—dominates the life of the community. Rituals add emotional impact. The latest techniques of applied psychology could not surpass the primal religions in keeping a body of religious thought alive without a written repository of religious information.

The Great Religions. Measured by their concern for humane values, many religions may be ranked highly. For Parsees, the salient virtues are liberality, justness, friendliness, and sincerity; the major duties are to make enemies into friends, to change the wicked into the righteous, and to turn the ignorant into the learned. Confucius taught similar virtues: wisdom, humanity, uprightness, decorum, and truth. The principal Confucian duties are honesty, self-control, and the practice of the golden rule—Do not do to others what you would not wish done to yourself. For Parsees and Confucians the good life does not carry the additional burden of obligation to convert the unbeliever.

In the competition for religious survival, the number of people adhering to a religion has some significance, and the greatest number follow Christianity, Islam, and Buddhism. Two other religions merit special consideration because they were progenitors—Judaism (for Christianity and Islam) and Hinduism (for Buddhism). Hinduism and Judaism began some 4,000 years ago; Buddhism, Christianity, and Islam developed later. While Hinduism, Buddhism, Judaism, Christianity, and Islam have been termed the great religions, they cannot be shown to reveal any deeper insights than other religions. The features introduced by the great religions reflected the needs of communities that were growing big and complex.

HINDUISM STARTED WITH traditions practiced along the valley of the Indus River. The religion was based upon sacred scriptures called

Vedas; the seminal sources were the *Ramayana* and the *Mahabharata*. A section of the latter, the *Bhagavad-Gita,* was written with particular force: "I am become death, the shatterer of worlds; waiting that hour that ripens to their doom." Hinduism gave firm social stabilization to a turbulent community by assigning a clear role and status to every individual—the caste system. Caste had unshakable permanence, but people could still hope for a better future through reincarnation.

Judaism started with Abraham who, through his promotion of monotheism, simplified what had been a chaotic background of multiple gods, multiple cults, and multiple conflicts. There had been earlier moves toward monotheism, for example by King Amenhotep IV of Egypt, but they had not lasted. Judaism was different because it also entailed a pledge with God. "Abraham's covenant with God, being personal, did not reach the sophistication of Moses' covenant on behalf of an entire people. But the essentials are already there: a contract of obedience in return for special favor, implying for the first time in history the existence of an ethical God who acts as a kind of benign constitutional monarch bound by his own righteous agreements"(2).

FROM 500 B.C. to A.D. 600 a new dynamic stream of ideas caught the attention of religious thinkers; a kind of religious renaissance occurred in what has been termed the Axial Age. The unifying concept of the Axial Age was that while our world is constrained by human nature, physical reality, and injustice, there is a world *beyond* which is better than anything that we have experienced. It offers eternal peace and serenity to all those who embrace the "true faith." Of course, the concept of life after death was not new; it had obsessed the Egyptian upper classes. The Axial Age, however, brought universality—an afterlife became available to everyone who lived a good life, and this novel idea of "equal rights" to enter heaven spread like a forest fire from one culture to another. Unfortunately, different religions held different views on what constituted the good life.

Contributions to the Axial Age came from Israel, Greece, Egypt, Syria, Iran, Iraq, Arabia, India, and China. Much has been written about this explosive stage of religious development. "A new type of elite ... included the Jewish Prophets and Priests, the Greek Philosophers and Sophists, the Chinese Literati, the Hindu Brahmins, the Buddhist Sangha, and the Islamic Ulema." The "new scholar class" turned into "relatively autonomous partners in the major ruling coalitions and protest movements." (10). Some may argue that with the Axial Age we at last find reason playing a role in the content of religious faiths, and if pressed they might cite Socrates lending support to the notion of an afterlife. But although the background of the philosophers who debated immortality may have been impeccable, their conclusions could not be confirmed or refuted by observation, and sometimes their views were conspicuously self-serving. Socrates upheld the immortality of the soul, declaring that opposites are generated out of each other, so life must arise from death, and the soul must be released by the body. He presented these arguments the night before his execution—perhaps we can excuse the conflict of interest when he stated, "No one who has not studied philosophy and who is not entirely pure at the time of his departure is allowed to enter the company of the gods, but the lover of knowledge only" (11). Even religions such as Christianity, which upheld most of the arguments of Socrates, must have balked at the philosophical credentials necessary for salvation. Philosophers, like everyone else, take a leap of faith if they are to embrace a religion.

The religious backdrop to the Axial Age was complicated and confusing—the new dogmas brought a simpler religious vision and, as a more practical contribution, social reform. Were the teachings of tolerance by Buddha, Christ, and Muhammad religious or political? Surely the notion that the sick, the poor, the racial minority and the moral outcasts deserve care and sympathy is both religious and political. If we explore the liberal social underpinning of the changes ushered in by the Axial Age we can find a rational justification for the trends. The new religions were open to all; they gained strength by allowing believers of every rank to enjoy the privileges of religious participation and social acceptance.

Why did so much change take place during the millennium of the Axial Age? At this time, populations were enlarging rapidly and as the size and complexity of communities grew, the great religions had to cope with new problems. Initially, Hinduism and Judaism dealt with the social tensions in their expanding cultures by placing people into different categories. Divisions such as conquerors and conquered, rich and poor, led the Hindus to adopt a caste system. Within Judaism, the Pharisees separated people into good and bad. The emergence of Buddhism, Christianity, and Islam was driven by a wave of reaction against these divisive social attitudes.

So reason played a role in the birth of the Axial Age religions, and it continues to contribute to their success in appealing to large communities. Reason supported the social reforms underlying the new religions. Reason also, as always, assisted in the operation of the institutions responsible for religious practices. Yet there is no consensus on which of the great religions is the best, and there is not even a consensus on the criteria for making such a judgment. S. G. F. Brandon summarizes what a rational scrutiny of the world's religions tells us:

> The evidence of mankind's religions tells us primarily, and tells us the more certainly, about man himself—about his intimations, about his fears and his aspirations concerning his own status and destiny in the world of his experience. The information which we are thereby afforded will undoubtedly appear very prosaic in comparison to that which it would seem may be won by investigating man's concepts of deity. Nevertheless, such information has its value (5).

ENHANCED MENTAL STATES

Religious faiths promote enhanced states of mind. Rudolf Otto vividly describes this universal feature of religion:

> The feeling of it may at times come sweeping like a gentle tide, pervading the mind with a tranquil mood of deepest worship. . . . It may burst in sudden eruption up from the depths of the soul with spasms and convulsions, or lead to the strangest excitements, to intoxicated frenzy, to transport and to ecstasy. It has its wild and demonic forms and can sink to an almost grisly horror and shuddering . . . and again it may be developed into something beautiful and pure and glorious (12).

A Mexican descendant of the Toltecs testifies to the intensity of religious experience:

> When you reach the point that you can concentrate with all your will, inside there, you reach a point where you feel ecstasy. It's a very beautiful thing, and everything is light. Everything is vibrating with very small signals, like waves of music, very smooth. Everything shines with a blue light. And you feel a sweetness. Everything is covered with the sweetness, and there is peace. It's a sensation like an orgasm, but it can last a long time (13).

Such descriptions reveal the level of passionate exhilaration that can be attained through religion. Religious faiths give extreme happiness to the few who can achieve the highest emotional response, and less intense but still worthwhile contentment to the many who find a more serene, low-key pleasure in joining each other to celebrate traditional ceremonies.

Religious rituals offer a rational way of making enhanced mental states available to as many people as possible. Rituals lead the congregation away from the mundane world into spiritual fulfillment where day-to-day problems are left behind. Through a rational process of observation, generalization, and prediction, religious institutions have learned that enhanced states of mind can be attained by gathering people into congregations, with ceremonial singing, or dancing, while priests give encouragement, in places of worship which themselves inspire awe, where emotions can be shared—joy, pride, fear, guilt, or shame—in the imminent presence of supernatural forces.

Enhanced mental states have alternative names according to their contexts. In the general literature "ecstasy" suffices. In religious texts, they are often called "mystical experiences." In psychological writings, "peak experiences" are equivalent. In 1984 the president of the American Medico-Psychological Association coined the phrase "Cosmic Consciousness." All of these terms seem to describe the same thing.

Enhanced mental states are not easy to define; they can more readily be communicated by examples. William James wrote a classic work on religious phenomena, *The Varieties of Religious Experience,* based upon a series of lectures he gave in Edinburgh from 1901 to 1902. His first chapter is titled "Religion and Neurology." James shows how most of us encounter episodes of enhanced consciousness at some time in our lives. He quotes Charles Kingsley describing a common type of special awareness. "When I walk in the fields, I am oppressed now and then with an innate feeling that everything I see has a meaning, if I could but understand it. And this feeling of being surrounded with truths which I cannot grasp, amounts to indescribable awe sometimes. . . . Have you not felt that your real soul was imperceptible to your mental vision, except at a few hallowed moments?" *(14).* Arthur Koestler sheds further light on the subject: "because the experience is inarticulate, has no sensory shape, color or words, it lends itself to transcription in many forms, including visions of the cross, or of the goddess Kali; they are like dreams of a person born blind. . . . Thus a genuine mystic experience may mediate a *bona fide* conversion to practically any creed, Christianity, Buddhism or Fire-worship" *(15)*.

James analyzes such phenomena with considerable care. "We instinctively recoil from seeing an object to which our emotions and affections are committed handled by the intellect as any other object is handled. The first thing the intellect does with an object is class it with something else. But any object that is infinitely important to us and awakens our devotion feels to us also as if it must be *sui generis* and unique." Later, James argues that we value states of mind for only two reasons: "because we take an immediate delight in them; or else it is because we believe them to bring us good consequential fruits for life."

W. N. Pahnke has summarized the universal characteristics of enhanced mental states in contemporary terminology:

1. *Unity* of "self" with the "inner" and "outer" world.
2. *Transcendence* of time and space, into eternity.
3. *Elation* interpreted as joy and peace.
4. *Awe,* with a feeling of sacredness or wonder.
5. Conviction of ultimate *reality* gained through intuition.
6. *Paradox* because of inconsistencies that appear when descriptions of the experience are analyzed subsequently.
7. *Transiency* of the state compared to the more protracted time course of normal experiences.
8. Persistence of the *impact* of the experience on personality, understanding, or behavior *(16)*.

The key words here, taken together, amount to an unusual formula: *unity, transcendence, elation, awe, reality, paradox, transiency, impact.* Abraham Maslow provides a similar definition, but adds further components such as *perfection, effortlessness, self-sufficiency, order, inevitability, and finality (4)*. Enhanced mental states are generally pleasant; people therefore want to have them repeatedly. Religion promotes enhanced experiences, and enhanced experiences promote religion.

WHAT IS KNOWN about ecstatic mental states? A wide variety of agents and circumstances can induce them. Physiological stress resulting from inadequate sleep, physical exhaustion, pain, hunger, or thirst can predispose to changes in consciousness such as dreamlike experiences and increased suggestibility. In "brainwashing" technology, physiological deprivations are deliberately imposed to alter the mental state and facilitate the introduction of new ideas. Commonly abused drugs such as alcohol, cannabis, cocaine, and other narcotics are all capable of distorting our sense of reality in ways that fall into the category of altered states of consciousness. Again, James is eloquent: "One of the charms of drunkenness, unquestionably lies in the deepening sense of reality and truth which is gained therein. In whatever light things may then appear to us, they seem more utterly what they are, more 'utterly utter' than when we are sober" *(14).*

For a few dedicated individuals the transcendent revelations of religion can be reached by concentrated meditation, using the techniques of yoga and Zen. More often, religions ignite ecstatic experiences in congregations. Rita Dove describes a scene from her Church of African Methodist Episcopal Zion:

> I watched the older women of the Church "get happy"; I could see them gathering steam, pushing out the seams of their composure until it dropped down, the Holy Spirit falling upon them like a hatchet from heaven. Instead of crumbling they rose up, incandescent, to perform amazing feats—they tightroped the backs of pews, skipped along the aisles, threw off ushers and a half-dozen able bodied men with every shout. . . . A woman "full of the spirit" was indomitable; one could almost see sun-

beams dancing off the breastplate of righteousness, the white wings twinkling on the sandals of faith. And when it was over, they were not diminished but serene, as if they'd been given a tonic *(17)*.

Psychological stimulation within a congregation can have a powerful emotional impact, but physical stimulation can also be effective. In medieval Europe flagellation was a common religious practice, and it persists. K. L. Reichelt offers a vivid image of modern Buddhist monks beating each other over the shoulders with wooden slats. "A more fantastic scene . . . could hardly be imagined, especially at night time, when hundreds of men in grey gowns, beneath the dim light of the kerosene lamps, are racing around at high speed with dust whirling around them" *(18)*.

So, psychological and physical techniques have been rationally used by religions to help people attain enhanced mental states. Inhalation of smoke (and perhaps an associated lack of oxygen) also works. G. Schüttler describes a Tibetan oracle-priest who first inhaled smoke from a censer for some 15–20 minutes *(19)*. His eyes then rolled up and his face turned red, then gray. He began to groan, and then shriek. His utterances took on a staccato quality, then subsided into hissing. His chest would rock, his arms shot out as if he was warding off unseen objects, and his feet began to beat a rhythm on the floor. His respiration became shallow, rapid, and irregular; he glistened with sweat, and he drooled saliva. Then he spoke to the congregation. The scene has been reenacted by religious groups throughout history. In Siberia, Wenceslas Sieroszewski recounts how

the shaman stares into the fire on the hearth; he yawns, hiccups spasmodically, from time to time he is shaken by nervous tremors. He puts on his shamanic costume and begins to smoke. Soon afterward his face grows pale, his head falls on his breast,

his eyes half close. . . . Suddenly a succession of shrill cries, pierc-
ing as the screech of steel, sounds from no one knows where.*

RELIGION'S ANSWERS

The promotion of ecstatic or serene experiences is only one facet of
how religion affects people's lives. More important in most modern
cultures is the ability of religion to answer the imponderable questions about our purpose in the universe and beyond. Religion offers
answers, but the answers are not accessible to confirmation or refutation by observation. Belief is an act of faith that must come from
the attitudes we have acquired. Religion owes its resilience to the
needs it fulfills. In his book *On Human Nature*, Edward O. Wilson
compares religion with science. He concludes:

> But religion itself will endure for a long time as a vital force in so-
> ciety. Like the mythical giant Anteus who drew energy from this
> mother, the earth, religion cannot be defeated by those who
> merely cast it down. The spiritual weakness of scientific natural-
> ism is due to the fact that it has no such primal source of power.
> While explaining the biological sources of religious emotional
> strength, it is unable in its present form to draw on them, because
> the evolutionary epic denies immortality to the individual and di-
> vine privilege to the society, and it suggests only an existential
> meaning for the human species. Humanists will never enjoy the
> hot pleasures of spiritual conversion and self-surrender; scientists
> cannot in all honesty serve as priests *(20)*.

Religion and reason have been in conflict at times, but many of the
greatest experts in the use of reason have upheld religious convictions. Socrates embraced the religious views of his time, which fore-

---

* "Du Chamanisme d'après les croyances des Yakoutes" (1902), quoted by Mircea Eliade
in *Shamanism* (New York: Pantheon Books, 1964).

shadowed Christianity. Plato recorded what Socrates said about divine judgment in the year 399 B.C., and the key points have a familiar ring. According to Socrates, the death of our bodies is followed by the migration of our souls, so that our souls may

> have sentence passed upon them, as they have lived piously or not. And those who appear to have lived neither well nor ill, go to the river Acheron, and embarking on any vessels which they may find, are carried in them to the lake, and there they dwell and are purified of their evil deeds, and having suffered the penalty of the wrongs they have done to others, they are absolved, and receive the rewards of their good deeds, each of them according to their deserts *(11)*.

Francis Bacon* asserted that "a little philosophy inclineth man's mind to atheism, but depth in philosophy bringeth men's minds about to religion" *(21)*. Isaac Newton told how the wonder of science reinforced his belief in a God: "Whence is it that Nature does nothing in vain: and whence arises all that order and beauty that we see in the world? . . . does it not appear from phenomena that there is a Being incorporeal, living, intelligent, omnipresent, who is infinite in space . . . ?" *(22)*. In the twentieth century another great physicist, Max Planck, gave a lecture entitled "Religion and Science," distilling his thoughts on how science and religion come together: "It is the steady, ongoing, never-slackening fight. . . . against unbelief and superstition, which religion and science wage together. The directing watchword in this struggle runs from the remotest past to the distant

---

* Francis Bacon was well connected. He was the son of the Lord Keeper of the Great Seal, and his aunt was the wife of Sir William Cecil (Lord Burleigh). Bacon was later appointed to his father's job, and in due course he became Lord Chancellor. In a scandal reminiscent of present-day political corruption, he lost the post for allegedly accepting bribes. Considering his contribution to the philosophy of scientific method, it is curious that he never discussed William Harvey's discovery of the circulation of the blood, although Harvey was his personal physician (S. G. Blaxland Stubbs and E. W. Bligh, *Sixty Centuries of Health and Physick*. London: Sampson Low, Marston, 1931).

future: 'On to God!' " Plank argued that we should identify the world order of science with the God of religion: "but for the religious man God stands at the beginning, for the scientist at the end of all thinking."* Albert Einstein's religious thoughts changed over the course of his life. At elementary school he received Jewish instruction; he "strictly adhered to ritual precepts, and in consequence no longer ate pork. He even composed a few hymns to the greater glory of God, which he sang with great fervor at home and when he was walking in the street." As he entered adolescence he read science extensively, and "the result was downright fanatical free thinking." Later, he formulated a rather personal religious position: "To know that what is impenetrable to us really exists, manifesting itself as the highest wisdom and the most radiant beauty which our dull faculties can comprehend only in their most primitive forms—this knowledge, this feeling is at the center of true religiousness."†

Some scientists have concluded that the domain of science is confined to "how" issues, while religion deals with "why" issues, so there is no inconsistency in accepting both. Other scientists have reached a level of understanding which, they claim, enables them to recognize that they do not truly know anything because ultimate truths are unknowable. These views all reflect the humility of minds dedicated to expanding human knowledge without prejudice. They lay to rest the image of the scientist as disputing the existence of anything he cannot observe and measure.

SOONER OR LATER everyone must come face to face with anguish and terror—it is at such moments that religion is indispensable. How

---

* This lecture called forth "wild applause." The quotations come from J. L. Heilbron's *The Dilemma of an Upright Man: Max Planck as a Spokesman for German Science* (Berkeley: University of California Press, 1986).

† The quotations on Einstein's early religious views come from the English translation of Albrecht Fölsing's recent biography *Albert Einstein* (New York: Penguin Books, 1997). The last quotation is from Einstein himself, *The World as I See It* (1934) (New York: Wisdom Library, 1979).

does one face death without religion? Could anyone without faith experience the feelings conveyed by the warm, supreme reassurance: "Yea though I walk through the valley of the shadow of death, I shall fear no evil: for Thou art with me; thy rod and thy staff they comfort me"? If rational beings from another planet were to visit us, we would expect them to bring their own religion, with texts setting out their beliefs, descriptions of their deities, stories about their ancestors, and edicts on their morality. If the strangers had no religion, and they learned how many and varied ours are, they would conduct a survey to find which faith would best meet their particular needs.

9

REASON AGAINST RELIGION

*Men never do evil so completely as when they do it from
religious conviction.*

—BLAISE PASCAL*

THE PLACE OF reason in a critique of religion is like that of a
lawyer in a trial. In the direct examination, reason can explore and
justify the goodness of religion, but in the cross-examination it can
wield a hatchet to expose a record of inconsistency, bigotry, and oppression. In the eighteenth century David Hume saw priests as me-
diating "any practice, however absurd and frivolous which either

---

* *Pensées* (1670). A recent English translation by M. Turnell was published as *Pascal's Pensées* (New York: Harper, 1962). Pascal set down many different opinions, including strong
criticisms of religion such as this, and also strong endorsements. He struggled with the
choice between dogma and skepticism. "Either a God exists or He does not. There is no
middle ground. Reason is impotent to help us to decide. A game of infinite consequence
is being played in which heads or tails must win. We must gamble. We have no choice. Not
to wager that God exists is to wager that He does not exist. Which side will we take?" Pascal argues that in this gamble, Roman Catholicism is the best bet because it offers eternal
life, versus the uncertainty of agnosticism. But why did he confine the choice to the
Catholic God? He did not deal with the Protestant God, or the Islamic God, the Jewish
God, or the Hindu gods. So how does one choose between all the gods of all the religions?
This is a serious problem that weakens the argument to such an extent that it loses its co-
gency. Nevertheless, Pascal was an innovator who will be remembered for his great contributions to mathematics.

folly or knavery recommends to a blind or terrified credulity" *(1)*. In the nineteenth century Ambrose Bierce described religion as "a daughter of Hope and Fear, explaining to Ignorance the nature of the Unknowable" *(2)*. In the twentieth century H. L. Mencken was equally skeptical: "1) The cosmos is a gigantic fly-wheel.... 2) Man is a sick fly taking a dizzy ride on it. 3) Religion is the theory that the wheel was designed and set spinning to give him the ride" *(3)*.

A recurrent theme in this book has been the search to find some biological roots of human behavior in animals. When it comes to religion, the search is barren—there was more than a touch of whimsy in the voice of Aldous Huxley when he addressed this issue:

> Man is so intelligent that he feels impelled to invent theories to account for what happens in the world. Unfortunately, he is not quite intelligent enough, in most cases, to find correct explanations. So that when he acts on his theories, he behaves very often like a lunatic. Thus, no animal is clever enough, when there is a drought, to imagine that the rain is being withheld by evil spirits, or as a punishment for its transgressions. Therefore you never see animals going through the absurd and often horrible fooleries of magic and religion. No horse, for example, would kill one of its foals in order to make the wind change its direction. Dogs do not ritually urinate in the hope of persuading heaven to do the same and send down rain. Asses do no bray a liturgy to cloudless skies. Nor do cats attempt, by abstinence from cats' meat, to wheedle the feline spirits into benevolence. Only man behaves with such gratuitous folly *(4)*.

Beliefs in God and a life after death are detached from reason because they can only be known through authority, intuition, inspiration, or revelation. But reason can weigh the consequences of religion, so with this in mind we shall take a journey from the past into the present, looking at the evidence.

HUMAN SACRIFICES

The Aztecs excelled in emotionally charged ritual. Like many others, they gave their priests responsibility for sustaining the universe in general and their culture in particular. Like many others, they held that sacrifice was the best way to satisfy their gods. What made the Aztec priests unusual was their commitment to fresh human blood—beating hearts were the chosen gifts for the gods. The conspicuous cruelty of cutting the heart out of a conscious victim merits special examination: this is surely a situation where universal standards of behavior are violated and religion is discredited. What, exactly, took place?

The Maya flourished in Mesoamerica from about A.D. 300 to 950. They were magnificent architects, engineers, builders, artists, astronomers and mathematicians. Aztecs followed the Maya until the Spanish arrived in the sixteenth century. The conquistadors marveled at the Aztec capital, Tenochtitlán, with phrases such as "things never before seen or dreamed about." Hernán Cortés described how the government controlled a marketplace:

> There exists in this great square a large building like an audience hall, where ten or twelve persons are always seated and who act as judges and who give sentence on all cases and questions arising in this market, and who order punishment for those who break the law. And in the same square there are other people who continuously walk among the people, observing what is sold and the measures with which it is measured; and we saw one measure broken which was false.*

---

* This quotation is from a letter Cortés sent to the king of Spain, translated by A. R. Pagden in *Hernán Cortés: Letters from Mexico* (New York: Grossman, 1971). Cortés led the Spanish expeditionary force that conquered Mexico and he was well received by Montezuma in Tenochtitlán. Later Montezuma was abducted and forced into public submission. In the face of harsh Spanish repression, the Aztecs revolted and forced Cortés out of the city. In 1521 he destroyed it, and laid the foundations of Mexico City in its place.

Evidently the marketplace contained a department of justice, a securities and exchange commission, a bureau of standards, and a federal bureau of investigation.

But in addition to its highly organized and rational government, the Aztec Empire had a powerful religion, and its priests embarked on large-scale programs of human sacrifice. The selection criteria for suitable heart donors was quite challenging: "he was not fat, he was not big bellied, he was not of protruding navel, he was not of hatchet shaped navel, he was not of wrinkled stomach, he was not of hatchet shaped buttocks, he was not of flabby buttocks" (5). Most of the victims were prisoners of war. Four priests held each victim down on a sacrificial stone while a fifth cut the chest open and removed the heart. At the dedication ceremony of the main Aztec temple in Tenochtitlán, 20,000 prisoners were reportedly killed over four days—though this figure is hard to reconcile with the resources known to have been available for performing ritual massacre. The usual rate of sacrifice was about 15,000 hearts a year. Occasionally local young men, women, and children were included.

Holy Wars

How does one religion react to the claims of others? If we hold that our own faith is the only true one, reason tells us that other religions must be untrue. Our religion—more precisely our religious bureaucracy—will then use reason to formulate an appropriate attitude to other religions. The response may range from calling for a holy war to developing a position of tolerance. Dealing with other religions is but one example of the more general question: how should a religion react to *any* discordant ideas? An orthodox religious view of the world that has answers to all questions must encounter difficulty as new information confronts the established order. We shall now see how religious institutions have met the challenges thrown up by changing circumstances.

Certain religions, notably Christianity and Islam, have engaged in contests of strength that have been needless and disastrous—beyond all reason. The Crusades are the best-documented holy wars. Terry Jones and Alan Ereira describe how the eleventh-century Syrian poet Abu 'l-'Ala al-Maarri lived in a town called Maarrat an-Numan that was invaded by cannibals. "These cannibals were Christians who had marched three thousand miles on their way to save the Holy Land in the name of Jesus." Al-Maarri wrote:

> The world is divided into two sects:
> Those with religion but no brains
> And those with brains but no religion *(6)*.

There may have been undercurrents of political advantages gained by diverting the warlords of Europe to an external enemy, but the overt decisive force driving the Crusades was religious fervor. The Crusades were consecrated military campaigns. The first appeal for armed volunteers was made by Pope Urban II in 1095, at a meeting in Clermont, southern France. In his recruiting speech, Urban II said that the reward for all those "taking the Cross" was immediate remission of sins. "This I grant to all who go, through the power of God with which I am invested" *(7)*. Judging from the subsequent behavior of many Crusaders, this dispensation was much needed.

The clearest statement of "war aims" for the Crusades was made by Sinibaldo dei Fieschi, Pope Innocent IV (1243–1254),* who started from the premise that as God's vicar on earth, the pope has dominion over all things. Innocent formulated and promulgated a series of principles. First, the pope is empowered to punish anyone who breaks a natural law (and the pope decides what is natural). Second,

---

* Innocent IV was the son of a count. He was embroiled in political struggles for most of his papacy. He was trained in law, and he used his training to defend the legality of the Crusades. He was clear but not entirely persuasive. Innocent proclaimed that he could insist that Islamic countries allow Christian missionaries to enter and preach the Gospel, but there was no reciprocity, "for we must not equate them with us, as they are in error and we are on the way of truth." B. Z. Kedar, *Crusade and Mission* (Princeton: Princeton University Press, 1984).

the pope has the right to demand that non-Christian rulers should admit preachers of the Gospel into their countries. Third, the pope is entitled to take whatever steps might be necessary to protect Christians in any part of the world. Fourth, if non-Christian rulers oppress their Christian subjects, the pope has the authority to depose the rulers. Fifth, the pope can reclaim any territory that had once been under Christian rule. Sixth, the pope may also lay claim to land that had not hitherto been under Christian rule.

Stephen Neill describes the corresponding position that was taken by Islam*(8)*. Their religion was "a word for all men, and not only for the Muslim; all are challenged to obey. The Muslim ruler is, therefore, fully entitled to send out to the Christian sovereign the demand that he should abandon his unbelief and accept the Muslim creed; if he fails to do so, he has declared himself an enemy. . . . Those who resist may rightly be slain"* *(13)*.

These implacable exhortations fueled military expeditions that became nightmares of destruction. Countless thousands died and nothing was accomplished. Civilian communities of Christians, Muslims, and Jews were caught in the savage onslaught; Tancred, a Norman Crusader, reported: "Our troops boiled pagan adults in cooking pots; they impaled children on spits and devoured them grilled"*(6)*.

Militant Christianity established elite orders of knighthood, soldiers who professed religious vows and accepted high levels of disci-

---

* Arun Shourie sheds light on the spirit of Islamic Holy war, or *Jihad,* in his book *The World of Fatwas* (Delhi: HarperCollins, 1997). "The one who dies while waging *Jihad* or subsequently of any injury sustained in *Jihad* is a martyr and is guaranteed Paradise. Every martyr acquires the power to intercede with Allah for up to seventy of his relatives. . . . As the duty is an overriding one all means are permissible. 'War is stratagem,' the Prophet says. 'War is deceit.' Thus one may lie, one may kill the enemy while he is asleep, one may kill him by tricking him." Since heaven is difficult to reach, *Jihad* carries strong motivation. Those who are normally most likely to dissuade a young man from risking his life in war, his mother, wife, and children, benefit from encouraging him to fight, because they will be among the "seventy relatives." It is hard to conceive of more powerful inducements to kill and die in the name of religion.

pline to fight for their faith. The Knights Templar, the Knights of St. John of Jerusalem, and the Teutonic Knights all arose in this way. In the words of Barbara Tuchman:

> Failure seems to have taught them nothing. Like human lemmings each generation of Crusaders flung themselves into the fatal foot-steps of their fathers. Palestine itself, the battleground and the prize, became a second country if not a graveyard for half the families of Europe. Saint Bernard of Clairvaux, who preached the second Crusade, boasted that he left but one man in Europe to console every seven widows (9).

While Christianity has abandoned holy wars, Islamic fundamentalists have not. Osama bin Laden, kingpin of the International Islamic Front for Jihad, declares: "Hostility against America is a religious duty, and we hope to be rewarded for it by God."*

THE INQUISITION

The Roman Catholic church has grappled with new ideas in a variety of ways. The first approach was to entrench religious faith so deeply that any novel view of the world would be unnecessary. The sun rises in the east and sets in the west because God ordains it; grass is green because this is decreed in heaven; children die of sickness because it is God's will. People have no questions to pursue because everything is explained.

This fortress of dogma, with ramparts of unassailable authority, remained impregnable for centuries. But commercial and military interests began to promote independent rational thought because they could not afford to do otherwise. Reason, in the form of science, gained momentum as its products became indispensable. Dissatisfaction with the rigidity of religion was bound to follow. Lecompte De Noüy explains how frustration built up:

---

*Reported by *Time*, January 11, 1999.

It was the logical consequence of the intolerance of the Church, which had hoped to consolidate its authority by smothering science under a bushel. This attitude alienated truth-loving, intelligent men, who by their devotion to freedom of thought, based on a sense of human dignity and on their horror of obscurantism, were necessarily opposed to a discipline that imposed the contempt of observed facts. . . . the Church progressively became a vast administrative machine, jealous of its prerogatives and its authority, convinced of its infallibility even in questions unrelated to dogma. When it encountered individuals who dared to think independently in certain fields, it often looked at them with a suspicious eye. . . . The clergy should have understood that all attacks on freedom of thought, when neither morals nor dogma were concerned, made enemies of the very people it most needed (10).

The problems for the church increased because in addition to the tide of new information concerning the outer world, issues began to arise concerning the inner world and, in particular, the church's monopoly on interpreting the Bible. In response to these mounting pressures the Inquisition was established, with the mission and authority to attack "subversion" in all possible ways. With its mandate and its power the Inquisition unleashed a reign of terror.

The origin of the Inquisition can be traced to St. Augustine, who played a decisive role in preparing the church's response to heresy. St. Augustine advocated punishment in the form of imprisonment, confiscation of goods, and exile. A heretic came to be regarded as an enemy of the human race. Legislation was enacted to prohibit the faithful from giving protection to heretics, and secular princes were ordered to assist the church or run the risk of excommunication. In 1224 the punishment for heretics was formally chosen to be execution by fire, a way of killing as gratuitously cruel as Aztec sacrifices. For a religion that preached forgiveness, the selection of death by fire was vindictive and hypocritical. The procedures for the medieval Inquisition were put in place by Pope Gregory IX between 1231 and 1235.

The Inquisition was driven by dogma: if enemies of the church do not recant or convert, you can still save their souls by burning their bodies. Given this irrational starting point, the Inquisition went to work with ruthless efficiency that could only be achieved by highly rational organization. In 1348, it burned 17,000 Jews in Erfurt, Strasbourg, and Bavaria. Intercession with the Inquisition was dangerous but individuals of moral courage still spoke out from time to time. The rector of the University of Treves was chief judge of the Electoral Court, notable for condemning countless witches to death. He decided that this practice was wrong so he began sparing them. In response, the Inquisition argued that since the rector was helping witches he must have sold himself to the Devil, so he too was executed.

By the sixteenth century science, founded on independent rational principles, was identified as an enemy by the Inquisition. The case that brought matters to a head was a conflict that would set the tone for future hostilities over 300 years. The church had affirmed the traditional view that the earth stood motionless at the center of the universe, with the sun, moon, and stars revolving around us. In ancient Greece the Pythagoreans had already questioned this thinking; they suggested that the earth rotated around its own axis every twenty-four hours and revolved around the sun every 365 days— indeed, they argued that these motions were responsible for the occurrence of our days, nights, and years. In 1543 Nicolas Copernicus confirmed these heliocentric conclusions in his classic work *De Revolutionibus Orbium Celestium (On the Revolution of the Heavenly Bodies).** In an effort to be diplomatic he dedicated his work to the

---

* Copernicus was the son of a Polish merchant. He was educated in mathematics, astronomy, Greek, law, medicine, and theology. He was an ecclesiastic, appointed as canon of Frauenburg. He completed *De Revolutionibus* in 1530 but delayed publication until 1543, just before his death, because he knew the church would object.

pope, and to be on the safe side, his publisher added a disclaimer stating that the earth's motion was only proposed as a hypothesis, and was not asserted to be a fact. These tactics worked initially, but as the truth of the Copernican hypothesis emerged, Christianity, both Catholic and Protestant, combined in opposition. For the Protestants, Martin Luther declared with crystalline clarity: "People give ear to an upstart astrologer who strove to show that the earth revolves, not the heavens or the firmament, the sun and the moon. Whoever wishes to appear clever must devise some new system, which of all systems is, of course, the very best."* For the Catholics, the Inquisition deliberated on the matter and then concluded: "That the earth is not the center, but revolves around the sun, is absurd, false in philosophy, and from a theological point of view at least, opposed to the true faith." Nevertheless, people with independent minds drew their own conclusions, sometimes with tragic personal consequences. In 1600, Giordano Bruno was burned at the stake for, among other heresies, agreeing with Copernicus.

One of the most distinguished intellectual figures of the time was the great physicist Galileo Galilei, and he gave his full support to Copernicus and Bruno. Galileo was ordered to appear before the Inquisition. On February 26, 1616, he was commanded to correct his mistakes. Perhaps with Bruno's fate in mind, Galileo promised that he would cast off the Copernican view and cease from writing or speaking of it.

In 1632 a new pope was installed, whom Galileo regarded as a friend. In this situation Galileo began to write again, but with more tact. In his *Dialogues of the Two Great World Systems* he presented both the traditional and the Copernican arguments. The church was incensed. A Jesuit, Father Melchoir Inchofer, led the attack: "The opinion of the earth's motion is of all heresies the most abominable,

---

* The quotations in this woeful story of bigotry come from Bertrand Russell's *Religion and Science* (15).

the most pernicious, the most scandalous; the immovability of the earth is thrice sacred; argument against the immortality of the soul, the existence of God, and the incarnation, should be tolerated sooner than an argument to prove that the earth moves." Galileo was summoned to appear before the Inquisition again, in spite of his failing health. He was imprisoned and threatened with torture if he did not recant. He had little choice. In public, on his knees, the Inquisition forced him to state, "I abjure, curse, and detest the said errors and heresies . . . and I swear that I will never more in future say or assert anything, verbally or in writing, which may give rise to a similar suspicion of me." He also swore to denounce any heretics who maintained that the earth moved. He was placed under house arrest; he was not allowed to see his family or friends. He wrote that the falsity of Copernicanism was beyond doubt, and it was disproved by the most solid argument from God. Galileo became blind and the pope refused him permission to return to Florence to visit his doctors. In these sad circumstances, he died on January 8, 1642.

A public funeral with great solemnity would normally have followed and the Grand Duke wished to erect a suitable commemorative mausoleum. Cardinal Francesci Barberini, writing on behalf of the Pope and the Inquisition, made it plain that such recognition was not appropriate for one who, although he had died a good Catholic, was nevertheless still doing penance for a serious offence. If there had to be a monument with an inscription it must avoid anything that could harm the reputation of the Inquisition *(11)*.

The church had crushed reason and flouted justice, but it had won a hollow victory which would return to haunt it. In 1992 Pope John Paul II rehabilitated Galileo; in his statement the pope explained: "Pastoral judgement which the Copernican theory required was difficult to make. . . . Let us say, in a general way, that the pastor ought to show a genuine boldness, avoiding the double trap of a hesitant attitude and of hasty judgement, both of which can cause considerable harm" *(11)*.

Some would regard this limited apology as too little, too late for Galileo. He was a founding father of modern science. Before him, physics had been an offshoot of Aristotelian philosophy. After him, the stage was set for Newton and Einstein. In addition to his contribution to the heliocentric theory, Galileo was the first to make serious astronomical use of the telescope. He invented the thermometer, and his analysis of the velocity and acceleration of falling bodies led to a new understanding of dynamics. His laws of motion had relevance to military projectiles that must have interested his patron, the Duke of Tuscany—perhaps this is what protected Galileo from the stake.

THE INQUISITION BECAME particularly powerful in Spain, which had previously been an unusually tolerant country. By the fifteenth century the Spanish Inquisition resolved to establish religious and racial "purity." The burning of Jews and Muslims became commonplace. Pope Sixtus denounced its savagery, but his protests were of no avail. Execution was not the only punishment meted out by the Inquisition. Vincent de Ferrer (subsequently made a saint) and Pablo de Santa Maria (a bishop) "decreed that Jews and Moors should wear distinguishing badges, be deprived of the right to hold office or possess title, and should not change their domicile. In addition they were excluded from various trades . . . were not allowed to eat, drink, bathe or even talk with Christians; and were forbidden to wear any but coarse clothes" *(12)*. The Inquisition became the "thought police," investigating whether those who professed Catholicism were harboring any doubts. They even hunted Catholic academics, such as Antonio de Nebrija, who complained: "Must I reject as false what appears to me in every way as clear, true and evident as light itself? What does this sort of slavery mean? What unjust domination when one is prevented from saying what one thinks, although to do so involves no slight or insult to religion."* A new inquisitor at Llerena was so zealous he provoked a letter of complaint to the king:

> We the relatives and friends of the prisoners in the cells of the Inquisition at Llerena kiss the royal hands of Your Highness and testify that the inquisitors of that province, together with their officials, have persecuted and persecute both the prisoners and ourselves with great hatred and enmity, and have carried out many irregularities in the procedure of imprisonment and trial, and have maltreated not only the said prisoners, but also their wives and children and property *(12)*.

The Inquisition proceeded without restraint, and the interrogators became trapped by their methods. They justified their position by arguing that God would not allow the innocent to be wrongly accused.

> The occasional suspect who defies all torture and refuses to talk is a threat to the whole system and must be forced by any means to submit.... this process almost guaranteed that no one who was accused could escape without confession, crippling or death—often death by burning alive. As Father Spee recounts: "A certain religious recently discussed the matter with several judges who had lighted many fires and asked them how an innocent person once arrested could escape; they were unable to answer" *(12)*.

This indictment reveals arrogance, hypocrisy, and, of course, irrationality. The argument that God would allow only the guilty to be accused can be restated in the logical form: all those accused are

---

* De Nebrija was a force for moderation who could speak out only because he had the protection of Ximénez, Archbishop of Toledo, and later a cardinal. De Nebrija wrote *Gramática Castellana* in 1492 and dedicated it to Queen Isabella I.

guilty; X is accused; therefore X is guilty. Thus the inquisitors cynically discredited their own judicial process.

The Spanish Inquisition came to a temporary halt in 1808. It was restored in 1813 and continued for a further twenty-one years.*

THE TERROR OF the Inquisition was also turned on witches. Who were witches? What did they do? How were they recognized? Witches were defined as practitioners of malicious magic. Magic and prayer both attempt to change the future, so how do they differ? Ancient texts do not give a clear answer, but with the arrival of Christianity the issue was formally addressed. The church decreed that prayer is a petition submitted to God with humility and in compliance with the prevailing religious and secular laws. In contrast, magic is independent of church and state, and operates without concern for moral or legal consequences.

While there may be overlap between the activities of scientists, physicians, and witches, the most important feature distinguishing witchcraft is its malicious nature. Some "target" must suffer as a consequence of witchcraft, but this criterion is imprecise, for medical or scientific failures can result in people being harmed. What if a scientific attempt to produce a flying machine ends in a crash, or a medical attempt to treat cholera ends in toxic reactions?

Further characterization of witches is available. Barbara Rosen summarizes the usual features of witches in the heyday of the Inquisition:

More women than men were called witches because witchcraft deals predominantly with the concerns of women and their world was a much more closed and mysterious society to men in the

---

* The Vatican recently assembled a panel of scholars to submit the deeds of the Inquisition to critical scrutiny—for the first time. Cardinal Roger Etchegaray said: "The church cannot cross the threshold of the new millennium without pressing its children to purify themselves in repentance for their errors, infidelity, incoherence" (Associated Press, October 30, 1998).

fifteenth century than it is now. Woman was regarded as deficient in the rational faculties—and since she was usually pregnant or nursing (the least intellectual of states) this is an understandable view. Her physical changes and functions were mysterious, particularly that of childbirth, which was assisted by women only, and about which doctors were astonishingly ignorant. . . . midwives were suspected of witchcraft more than any other women *(13)*.

Later, Rosen adds that "the practitioners of country magic were most often women. This simple fact was enough to release in the Inquisitors a violent mixture of antifeminism and sex-obsession." Certainly, there was an insistence that the witch cult entailed sexual intercourse with the Devil and with animals. Ecclesiastics debated whether the Devil was capable of producing offspring. From his research, H. R. Trevor-Roper found that the Fathers were assiduous in their inquiry, but divergent in their conclusions.

No detail escaped their learned scrutiny. As a lover, they established, the Devil was "of freezing coldness" to the touch; his embrace gave no pleasure—on the contrary, only pain; and certain items were lacking in his equipment. But there was no frigidity in the technical sense: his attentions were of formidable, even oppressive solidity. That he could generate on witches was agreed by some doctors (how else, asked the Catholic theologians, could the birth of Luther be explained?); but some others denied this, and others considered that only certain worm-like creatures, known in Germany as *Elben,* could issue from such unions. Moreover, there was considerable doubt whether the Devil's generative power was his own, as a Franciscan specialist maintained ("under correction from our Holy Mother Church"), or whether he, being neuter, operated with borrowed matter. A nice point of theology was here involved and much interested erudition was expended on it in cloistered solitudes *(14)*.

The imagination of experts in witchcraft was bizarre; they established that when witches were not involved in making love to the Devil, they passed their recreational time bringing impotence upon bridegrooms and suckling toads, bats, or weasels. Since witches were such expert practitioners of unusual sexual practices, it is perhaps surprising that tests for identifying them tended to have a spiritual context. For example, in one ordeal, the suspect witch was trussed and thrown into the water. If she sank, she was innocent; if she floated, she was guilty. Flotation indicated that the alleged witch had rejected her baptism, so the water would not receive her—a curious argument that twists reason to ensure that one way or another, the accused would be punished.

Scorn and suspicion were poured on witches. In a God-fearing society misfortune could come only from the malicious acts of "outsiders" hidden within the community. Witch hunting was gratuitously cruel and it was not confined to the Inquisition. King James I of England watched the interrogation of witches and he even invented a new technique of torture: after pulling off the fingernails, needles were inserted into the nail bed. James knew so much about witches that he wrote a book on demonology, in which he ventured to suggest that the Devil had experimented in the management of infertility by using semen obtained by massaging the testicles of corpses (14). English witch hunts continued into the nineteenth century; in 1863 an old man was lynched for being a wizard. Belief in evil spells continues to this day.

What sustained witch hunting for so long? Rosen points out the need to recognize

the desperation of lives played out in restricted surroundings and to feel the helplessness of people before a succession of natural catastrophes which they did not in the least understand. Precious brewings and bakings are lost; precious livestock suffer from inexplicable sicknesses; crippling diseases strike; and over and over again, children scream and suffer helplessly and die. There are very few literate people today who can enter into an existence in which one bears ten children and watches five of them die in infancy; but such an existence is at the basis of witch-belief *(13)*.

Religious institutions fostered the idea that witches were responsible for these events, and they taught that the solution was to kill witches.

Yet the alleged witches never posed a significant threat to the religions that persecuted them. The circumstances, bad as they were, could not explain the aggressive "search and destroy" policy. Perhaps the most persuasive interpretation of witch hunting is that by terrorizing scapegoats, religious institutions exercised social control.

Nonpractice of witchcraft is no safeguard against an accusation of witchcraft. How then do you protect yourself from false accusations; by avoiding quarrels; and by doing everything possible not to lose the support of your kin groups. Thus the occasional killing of a supposed sorcerer results in much more than the mere elimination of a few actual or potential antisocial individuals. The violent incidents convince everyone of the importance of not being mistaken for an evildoer. . . . people are made more amiable, cordial, generous, and willing to cooperate *(15)*.

In this way religious institutions reinforced their control over society, and incidentally protected themselves from being blamed for calamities. But by reinforcing stereotyped conformity within a culture—through the persecution of scapegoats—the religious institutions also stifled innovation.

CREATIONISM

In creationism, religion offers a present-day example of the dilemma that brought the Inquisition into conflict with science, the problem of dealing with newly discovered facts that are not in keeping with traditional religious teaching. Instead of adjusting to the changing circumstances, religion attempts to suppress reason. Instead of confining dogma to matters that are beyond verification, religion continues to sustain an inflexible and irrational position when there is evidence against it "beyond reasonable doubt." Creationism defies Galileo's knife, the rational dictum that observations take precedence over theories.

The Bible states that the origins of the universe and of *Homo sapiens* were separated by only six days. Archbishop Usher calculated, from the genealogies recorded in Genesis, that Creation took place in the year 4004 B.C. Dr. Lightfoot, of Cambridge University, added further precision to Archbishop Usher's calculation by determining that the origin of man occurred at 9 A.M. on October 23. Rational examination of the evidence refutes these claims. Furthermore, according to Genesis, animals did not die before Adam had eaten the forbidden fruit, yet the study of fossils has revealed numerous species that were extinct long before *Homo sapiens* appeared. Furthermore, the size of Noah's Ark has become a matter of wondrous conjecture since the discovery that it must have accommodated over a million species.

These are some of the problems that exercised the minds of theologians through the nineteenth century, until Charles Darwin published his theory of evolution by natural selection, setting the scene for a public debate that polarized science and religion.* Bishop Wilberforce led the attack with the slogan: "The principle of natural selection is absolutely incompatible with the word of God," and the battle has continued into the twentieth century. In 1925 John Scopes, a high school science teacher in Dayton, Tennessee, was arrested for teaching evolution. He was accused of violating an antievolution law that had been passed by the state legislature the year before. H. L. Mencken covered the events for the *Baltimore Sun*. Over the days leading up to the trial, the road to Chattanooga was lined with signs

---

* Charles Darwin was sent to Cambridge to study divinity in 1827, but he spent his time socializing, hunting, shooting, and learning science. Eventually he settled down in the village of Down, participating in local social life and attending church functions although he was an agnostic.

proclaiming "Sweethearts, Come to Jesus," "You Need God in Your Business," and "Prepare to Meet Thy God." The mood was one of a carnival. Two chimpanzees had been brought from a circus to testify for the prosecution. Mencken reported on the atmosphere in Chattanooga as the trial started:

> There is, it appears, a conspiracy of scientists afoot. Their purpose is to break down religion, propagate immorality, and so reduce mankind to the level of the brutes. They are the sworn and sinister agents of Beelzebub, who yearns to conquer the world, and has his eye especially on Tennessee.

The local population was in a state of high excitement. Mencken describes an incident he witnessed during a religious service. When the preaching stopped, a woman stood up and denounced the reading of books; she said that a wandering book agent had come to her cabin and tried to sell her a specimen of his wares: "She refused to touch it. Why, indeed, read a book? If what was in it was true then everything in it was already in the Bible. If it was false then reading it would imperil the soul." Another speaker added that education was a snare. "Once his children could read the Bible, he said, they had enough. Beyond lay only infidelity and damnation. Sin stalked the cities. Dayton itself was a Sodom."*

The Scopes trial laid bare the conflict between arguments based upon dogma (the Bible is the repository of all truth) and arguments based upon observation (the dating of fossils and the finding of tools and bones of early hominids). The Scopes trial resonated with overtones of the Galileo trial.

Not surprisingly, Scopes was convicted, but the Tennessee Supreme Court reversed the decision on a technicality. The law was

---

* A collection of the biting and entertaining journalism of H. L. Mencken, including the passages cited here, was put together in *The Impossible Mr. Mencken,* edited by Marion Elizabeth Rodgers (New York: Anchor Books, 1991).

not repealed until 1967. As recently as 1987, Louisiana had a Creationism Act that forbade anyone teaching the theory of evolution in public schools unless this was accompanied by instruction in what was called "creation science," a body of pseudo-information offering a quasi-religious alternative to evolution. This law naturally reflected the wishes of the majority of voters in the state. The controversy continues today among several North American school boards.

THE NEUROLOGY AND PSYCHIATRY OF RELIGION

We shall now reexamine mystical experiences to weigh the claim that in special individuals they are divine intercessions. Are mystical experiences natural or supernatural? As we look at these questions we must recognize that we cannot *disprove* the supernatural origin of a mystical experience by explaining it in terms of a neurological or psychiatric disturbance in the brain.* All we can do is offer a biological interpretation and apply Ockham's razor—the simplest explanation is the rational choice.

A. M. Ludwig defines an altered state of consciousness as "any mental state induced by physiological, psychological or pharmacological manoeuvres or agents, which can be recognized subjectively by the individual himself (or by an objective observer of the individual) as representing a sufficient deviation in subjective experience or psychological functioning from certain general norms for that individual during alert, waking consciousness" (15). To look into the nature of these phenomena, we must focus our attention on the cerebral cortex, the newly evolved part of the brain that forms a convoluted mantle covering the deeper, central regions. The cerebral cortex is in a continuous state of electrical and chemical flux. This activity is highly coordinated, and it is organized so that multiple tasks

---

* In their paper "The Neural Substrates of Religious Experience," J. L. Saver and J. Rabin go further: "Indeed, it has been argued that demonstrating the existence of a neural apparatus sustaining religious experience can reinforce belief because it provides evidence that a higher power has so constructed humans as to possess the capacity to experience the divine." *Journal of Neuropsychiatry and Clinical Neurosciences* 9 (1997):498–510.

can be carried out simultaneously without interfering with each other. The pattern of electrical activity has rhythmic components that vary according to the individual's level of wakefulness. This intricately balanced conglomerate of electrochemical interaction can be thrown into disarray. The commonest cause is excessive, paroxysmal discharge of high-voltage activity—the physiological basis of epilepsy. We do not know how epilepsy became associated with the supernatural, but we can guess. A major generalized convulsion (grand mal seizure) has a startling onset with sudden falling to the ground and loss of consciousness, followed by wild involuntary movements, and a dramatic ending with rapid spontaneous recovery. If demonic spirits could temporarily take possession of the human body, how else would they declare themselves?

Epilepsy is a unique disturbance of cerebral function. As an accident of nature, it gives us considerable insight into how the brain works. The abnormal electrical brain discharges can be precipitated by a variety of causes, such as stroke, tumor, trauma, or intoxication. Often we cannot find the cause for a patient's seizures. The result of the epileptic discharge will depend upon which part of the brain is involved. If the entire brain is affected, the attack is a generalized convulsion with loss of consciousness. If the abnormal electrical discharge is localized to a portion of the brain, the features will depend on the normal function of that particular portion. When the area concerned with movement of an arm is the focus of a seizure, the arm will start jerking. The link between cerebral localization and external manifestation is precise and consistent, so observation of an epileptic attack confined to one limb will allow a neurologist to pinpoint the site of abnormal function in the brain.

One common form of seizure leads to altered states of consciousness. Here, disturbed electrical activity usually affects the part of the brain beneath the temple—the temporal lobe. Such attacks are termed *temporal lobe epilepsy* or *complex partial seizures*. In addition to the psychological changes *during* temporal lobe attacks, there can be abnormal mental phenomena *between* attacks. In order to discuss these phenomena, we must introduce the relevant terminology. The seizure is the *ictus* (from Latin *icere,* to strike). Altered mental states during the seizure are *ictal;* altered mental states between seizures are *interictal.*

What are the typical features of temporal lobe epilepsy? When the abnormal electrical discharge is confined to the temporal lobe, the seizure causes alterations in the content of consciousness rather than loss of consciousness. Often there is an intense feeling of familiarity—this is called *déjà vu.* The opposite also occurs, an intense feeling of unfamiliarity—this is called *jamais vu.* There may be vivid visual disturbances with subsequent inability to recall the precise details. Objects may appear larger or smaller than normal. Patients can have auditory hallucinations, such as hearing their names called or sensing an indefinable melody. They may experience unusual smells or tastes, or peculiar abdominal sensations. They may report feeling a breath of wind, or someone brushing past them. They may describe the impression of being in a dream or of feeling detached. There can be emotional reactions, such as fear or sadness. Repetitive utterances can occur; these may involve meaningful words or they may be nonsense. There may be abnormal forms of motor activity, such as chewing, lip smacking, swallowing, smiling or laughter, turning the eyes or the head, scratching, rubbing the hands, or kicking. This wide range of possible epileptic phenomena does not pose a diagnostic problem, because the same pattern tends to be repeated consistently in each patient.

Occasional patients with temporal lobe seizures develop an *interictal* disorder termed the schizophrenia-like syndrome, or the schizophrenia-like psychosis of epilepsy *(17; 18),* in which hallucinations and delusions replace reality. This does not destroy the fabric of the mind—the personality remains intact.

Do any of the phenomena associated with temporal lobe epilepsy shed light on mystical experiences in normal people? The central question is whether altered states of consciousness are generated in the temporal lobes of normal brains, in addition to the abnormal brains of epileptics. In his *Varieties of Religious Experience* William James wrote: "A more pronounced step forward on the mystical ladder is found in an extremely frequent phenomenon, that sudden feeling, which sometimes sweeps over us, of having 'been here before,' as if at some indefinite past time, in just this place, with just these people, we were already saying just these things"*(19)*. Alfred Tennyson echoed this notion:

Moreover, something is or seems,
That touches me with mystic gleams,
Like glimpses of forgotten dreams.*

We would go too far if we were to suggest that all altered states of consciousness derived from temporal lobe epilepsy, but equally we would not go far enough if we were to ignore the neurological evidence. Under the heading "God in the Temporal Lobes," David Wulff develops a suggestion, put forward by Michael Persinger, that mystical experiences are produced in normal people by "microseizures" in the temporal lobes *(20; 21)*. According to this view, "microseizures" can be precipitated by a wide range of triggering stimuli, including the emotionally charged mental states evoked by religion. In order to avoid confusion with the seizures of epilepsy, we shall use the term "microsynchronization" instead of "microseizure" for these phenomena, which involve the simultaneous discharge of a circumscribed, tightly connected, small group of nerve cells in the brain.

---

* "The Two Voices" was published in 1842, though it was dated 1833. It is a poem of 462 lines; lines 379–381 make up the quotation. One voice starts the poem with a thought of suicide and the other argues against it:

A still small voice spake unto me,
"Thou art so full of misery,
Were it not better not to be?"

Then to the still small voice I said;
"Let me not cast in endless shade
What is so wonderfully made."

Microsynchronization, defined in this way, has *no tendency to spread or build up electrical intensity to create a seizure.*

Returning to consider real temporal lobe seizures as opposed to microsynchronization in normal temporal lobes—the neurological literature contains graphic accounts of patients having intense religious experiences. G. Sedman reports epileptics seeing the heavens open, hearing God speak, feeling themselves transfigured, and even sensing that they, themselves, have become God *(22).* K. Dewhurst and A. W. Beard examined the religious features of schizophrenia-like syndromes in twenty-six patients. Six out of the twenty-six underwent religious conversions. One was a bus conductor: "in the middle of collecting fares he was suddenly overcome with a feeling of bliss. He felt he was literally in heaven. . . . he said that he had seen God and that his wife and family would soon join him in heaven" *(22).* Another patient described his conversion in the following way: "It was a beautiful morning and God was with me and I was thanking God, I was talking to God; I was entering Aldwych, entering the Strand, between Kingsway and the Strand."

Can temporal lobe epilepsy explain any of the experiences reported by celebrated religious mystics? St. Paul fell down on the road to Damascus, experienced visual hallucinations, and converted to Christianity. The abnormal electrical activity of temporal lobe epilepsy sometimes spreads across the brain to result in a generalized convulsion with loss of consciousness; is this what happened to St. Paul? St. Teresa of Ávila experienced visions and attacks of unconsciousness—again, we have a possible diagnosis of temporal lobe epilepsy. St. Thérèse of Lisieux was another example; she had visual hallucinations, including mystical experiences, with "strange and violent tremblings all over her body." In all these cases, the difficulty in reaching firm conclusions is lack of information. But the story of St. Joan, the Maid of Orléans, merits special consideration because her experiences were recorded in detail as documentation of her trial.

Joan was born in 1412; she grew up as a peasant girl, at a time when a war was raging over the succession to the French throne. The Treaty of Troyes, signed by the father of the dauphin, assigned succession to the English royal line. The treaty was contested by the dauphin, who claimed the throne for himself, and tradition dictated that the king of France should be crowned in Reims, which was in the hands of the English. In this setting, Joan began to hear voices telling her to drive the English out of France and crown the dauphin in Reims cathedral. These were challenging assignments! Her first task was to raise the siege of Orléans; her next was the capture of Reims. She astonished everyone, except herself, by accomplishing everything her voices commanded, including the crowning of the dauphin. Then her military and political skills declined, and eventually she was captured by Burgundian allies of the English. She was tried for heresy by an ecclesiastic court; at the end of a long hearing she recanted, so instead of being executed she was sentenced to prison for life. Part of the evidence against her had been her habit of wearing men's clothes, and after the trial she was again found wearing them. She was retried and this time she was convicted and burned at the stake as a relapsed heretic in 1431. The verdict was reversed in 1456 and in 1920 Joan was made a saint.

From the record of her trial, it is clear that Joan had total confidence in her voices: "Everything that I have done that was good I did by command of my voices." They told her that she had a divine mission: "I was in my thirteenth year when God sent a voice to guide me. At first I was very much frightened. The voice came towards the hour of noon, in summer, in my father's garden. I had fasted the preceding day. I heard the voice on my right hand, in the direction of the church. I seldom hear it without a light. That light always appears on the side from which I hear the voice" (24). Joan also described abnormal olfactory experiences, saying that she had embraced St. Catherine and St. Margaret; she remembered smelling their perfume. Sometimes size was distorted in her visual hallucinations. She saw angels "in great multitudes and in very small dimensions." She attributed her transvestism to divine origins: "It is a little thing and of no importance. I did not don it by advice of the men of this world;

I donned it only by the command of God and the angels. . . . I do not think I do ill: And as soon as it please God to order this, then I shall cease to wear it" *(24)*. Joan's intelligence and integrity shine out in the record.

Her olfactory, visual, and auditory hallucinations are all in keeping with temporal lobe epilepsy, and the sense of distorted size provides further support for the diagnosis. It is interesting that she heard her first voice and saw her first vision after a fast, because a low concentration of glucose in the blood often precipitates seizures in susceptible subjects. So the various pieces of evidence, taken together, are certainly compatible with a diagnosis of temporal lobe epilepsy and the interictal schizophrenia-like syndrome. Could she have suffered from a hysterical disorder? Hysterical patients tend to be suggestible, but throughout the trial Joan was firm in her position and refused to be led by the court. For example, she was pressed to describe, from her visions, the clothes worn by St. Catherine and St. Margaret, but she was adamantly unable to recall them. These features make hysteria unlikely.

FINALLY, WE MUST consider psychosis as a possible cause of mystical experiences. Manic-depressive psychosis and schizophrenia lead to disordered thought, hallucinations, delusions, and disturbances of mood. However, in contrast to the schizophrenia-like syndrome associated with temporal lobe epilepsy, untreated psychoses have a grave prognosis. In manic-depressive psychosis the personality can undergo intermittent disruption, while in schizophrenia it simply disintegrates into madness.

The thoughts of psychotic patients are shaped by their culture. Medieval patients were fearful of magic and demons. The modern psychotic patient complains of being influenced by microwaves, lasers, or aliens from outer space. Israel is an exception where religion still plays a large a part in the culture and it still dominates the thought of many psychotics. L. Perez has described his experience with Jewish and Christian patients in Jerusalem. Psychosis did not generally take patients outside the cultural confines of their personal background. Jewish women identified themselves with such figures as the prophetess Deborah, or even the wife of the Messiah. Christian women usually perceived themselves as the mother of Christ. Sometimes there is a messianic mission; for example, one patient decided to start by healing all the other patients in his ward of the hospital. Another went directly to the Jerusalem Television Service and asked for at least two hours of live time so that he could bear his glad tidings as quickly as possible to all the people of Israel. The common denominator of these messianic patients was the combination of "a) A delusional system characterized by the belief in a redemptive task assigned to the patient by supernatural powers; b) hallucinations ascribed by the patient to those powers; and c) behavior consistent with the delusional system" (25).

Psychosis gives us an inkling of how culture can determine the content of altered states of consciousness, but it is unlikely that many of the celebrated religious mystics were psychotic. Prior to the modern era of psychopharmacology, treatment of psychoses was so inadequate that deterioration would have led to the label of lunacy— unless, of course, they died before this could happen.

10

ART

People who try to explain pictures
are usually barking up the wrong tree.
—PABLO PICASSO*

W HAT ROLE DOES reason play in art? Is reason embedded in
art in the same way that it is embedded in morality and science, or
does it have a lesser part to play, as in religion? A poem, an aria, or a
painting can transmit feelings directly and powerfully, and the impact of these feelings can approach the enhanced mental states that
we discussed as religious ecstasy.† Rebecca West writes of "this blaz-
ing jewel that I have at the bottom of my pocket, this crystalline concentration of glory, this deep and serene and intense emotion that I

---

* This quotation comes from R. Goldwater and M. Treves, *Artists on Art* (New York: Pantheon, 1945).
† J. E. H. MacDonald, a Canadian painter (one of the Group of Seven), gave a crisp definition of visual art: "A picture is not a reflection of a thing seen but a compound of feelings aroused in the artist by the things seen, resulting according to his skill in a more
concentrated impression than the natural objects can give." This quotation comes from
"Notes for paper on Relation of Poetry to Painting with special reference to Canadian
Painting, Oct 20, '29," J. E. H. MacDonald Papers, MG30 D111, vol. 3, file 3–20, National
Archives of Canada.

feel before the greatest works of art" *(1)*. In a similar vein, Ellen Dissanayake describes the aesthetic experience as:

> the swooping sensation inside the chest as one watches a dancer leap or a cathedral vault soar, or hears a gathering crescendo; the contraction of the throat or even trembling and weeping while one follows a complex harmonic progression; the sensation on the inside of one's hand of mentally caressing a sculpted curve of stone or wood; the wish somehow in some unimaginable way to gather a poem (its words, all it implies) and fasten it to oneself, blend with it, possess it; the unexplained feeling that words have shapes, that colors can be touched or heard, that sounds have contours and weight *(2)*.

In some sensitive individuals quite simple stimuli can generate similar experiences. Tennyson tells us how he could induce an enhanced mental state introspectively:

> A kind of waking trance—this for lack of a better word—I have frequently had, quite up from my boyhood, when I have been all alone. This has come upon me through repeating my own name to myself silently, till all at once, as it were out of the intensity of the consciousness of individuality, individuality seemed to dissolve and fade away into boundless being, and this not a confused state but the clearest, the surest of the surest, utterly beyond words—where death was an almost laughable impossibility—the loss of personality (if so it were) seeming no extinction, but the only true life. I am ashamed of my feeble description. Have I not said the state is utterly beyond words?*

---

* This appears in a letter to Mr. Blood, quoted by William James in his *Varieties of Religious Experience* (1902) (Boston: Harvard University Press, 1985).

These images of the aesthetic experience resonate with the images we have seen in exploring religious ecstasy—the language is almost identical. We have attributed religious mystical states with microsynchronization of activity in nerve cells. To be consistent, and to comply with Ockham's rational principle of choosing the simplest hypothesis, we may argue that the aesthetic experience is also generated through microsynchronization in the brain.

WHAT IS ART?

How do we define art? What determines whether art succeeds or fails in its role of communicating the aesthetic experience? How do we evaluate art? Attempts to answer these questions open a can of worms, for there is astonishing difficulty in dealing with what seem to be straightforward issues concerning an endeavor of great importance to all human societies. Roger Scruton illustrates the problems of defining art:

> Suppose an anthropologist, engaged in the study of a remote tribe, begins to classify its activities according to the purpose which they serve. He sees people tilling the fields, planting and harvesting, and attributes this activity to the desire for food. He observes the manufacture of clothing, tools and medicines; the building of houses and the rituals which are addressed to higher powers. But there are certain activities which seem to have no purpose. Members of the tribe take "time off" from urgent things, in order to build weird structures, to make wailing noises, to move in mesmerising dances, or to throw cakes of dung high into the treetops. Suppose that the tribesmen have a word for all these activities (though its application to each of them is disputed by at least one of their members). Suppose that, when the anthropologist asks what this word ("Schmart") means, and what it is that works of schmart have in common, people rack their heads and come up with some unilluminating technicality: the point of Schmart, they say, is that people have a schmaesthetic interest in it. Pressed fur-

ther, they begin to claim that other tribes, at other periods of history, also take a schmaesthetic interest, though not necessarily in quite the same things. One of their philosophers eventually proclaims that the schmaesthetic is a human universal. . . . The important thing, they say, is the *experience* of schmart *(3)*.

Scruton's allegory brings home the important questions about art, but it does not answer them. If art is defined as something that evokes aesthetic experiences, how is this achieved? How is art created, or, in the terminology of neurology, how do artists select and assemble the stimuli that will give people cerebral microsynchronization? From Scruton's dung-throwing tribe, it is clear there is a wide range of techniques that can be used to achieve the experience, and what works in one culture may not work in another. Over the course of history countless lives have been dedicated to producing art and to explaining it through philosophy and interpreting it through criticism. Philosophers have asked what it means to say that one work of art is better than another, and critics have helped people to appreciate art. Yet there is no consensus. As Joseph Machlis puts it:

> Art, like love, is easier to experience than define. It would not be easy to find two philosophers who would agree on a definition. We may say that art concerns itself with the communication of ideas and feelings by means of a sensuous medium—color, sound, bronze, marble, words. This medium is fashioned into works marked by beauty of design and coherence of form. They appeal to our mind, arouse our emotions, kindle our imagination, and enchant our senses.*

---

* *The Enjoyment of Music,* 3d ed. (New York: Norton, 1970), which grew out of the introductory course in music at Queens College of the City University of New York. The book provides an enchanting journey through the history of music and is notable for illustrating much of the evolution of musical style with visual art.

Machlis is eloquent, but his analysis really gives no more insight than the philosophers in Scruton's tribe.

ART IS MADE up of visual forms—drawings, paintings, sculpture and architecture; auditory forms—music, recitation, and literature; or the two forms combined—theater, opera, and dance.* Reason makes a vital contribution to the content of art. In the words of Leonardo da Vinci: "The painter who draws by practice and judgement of the eye without the use of reason is like the mirror that reproduces within itself all the objects which are set opposite to it without knowledge of the same"*(4)*. But reason also contributes to technique, for example in establishing the laws of perspective and the visual properties of light. Again Leonardo writes: "The first requisite of painting is that the bodies which it represents should appear in relief, and that the scenes which surround them with effects of distance should seem to enter into the plane in which the picture is produced by means of the three parts of perspective, namely the diminution in the distinctness of the form of bodies, the diminution in their size, and the diminution in their color"*(4)*.

In music, reason has led to an explanation of the physics of sound. Pythagoras (572–477 B.C.) showed that the interval of an octave is obtained by halving the length of a vibrating string. Mersenne (1588–1648) established the laws governing the pitch of a vibrating string by determining the relationship between length, stretching force, and mass *(5)*. Reason, through the technology of printing, allowed the pitch, rhythm, and interplay of multiple parts to be preserved and disseminated. Reason led Aristotle to define the formal qualities of a work of art as coherent unity combined with diversity, theme, and balance.†

---

* Gourmets might wish to add gustatory and olfactory forms—food and drink beyond mere sustenance.

† While the terminology is not consistent, these ideas were presented in the *Poetics*, written during the fourth century B.C.

QUESTIONS ABOUT THE nature of art and aesthetics have never been easy, and their difficulty has increased over the course of the present century. Two kinds of change have worked together to create deep problems, and reason contributed to each.

First, science and technology found ways of reproducing and storing images of the world that infringed on the traditional territory of artists. Photography generated cheap, realistic pictures, and the cinema—later television—brought the performing arts to everyone at low cost. Radio and sound recording allowed music to be played virtually everywhere. A demarcation dispute began to arise when artists producing "traditional culture" suddenly found themselves competing with new technological media producing "popular culture." The artists' elite position was threatened, and they responded by moving abruptly away from the conventions of their predecessors (6). Art has always been on the move, with new thoughts, new techniques, and new attitudes leading to the development of new schools. But the earlier process of evolution turned into revolution by the beginning of the twentieth century, as the modern and then postmodern movements introduced radical changes.* In the words of John Carey:

> the spread of literacy to the "masses" impelled intellectuals in the early twentieth century to produce a mode of culture (modernism) that the masses could not enjoy, so the new availability of culture through television and other popular media has driven intellectuals to evolve an anti-popular cultural mode that can reprocess all existing culture and take it out of the reach of the majority. This mode, variously called "post-structuralism" or "deconstruction" or just "theory," began in the 1960s with the work of Jacques Derrida, which attracted a large body of imitators among

---

* Many twentieth-century artists resented this interpretation and simply regarded the new direction of art as the culmination of everything that had gone before.

academics and literary students eager to identify themselves as the avant-garde *(6)*.

The second factor driving the shifts in artistic styles has been the impact of political egalitarianism.

The egalitarian argument for the new view can be summarized in the form of a logical deduction: all opinions of art are personal preferences; all personal preferences have equal weight; therefore all opinions of art have equal weight. Ellen Dissanayake gives a concise summary of the debate between adherents of classical Western "high art" and supporters of the contemporary postmodernist movement.

> The most up-to-date and influential practitioners and critics of the arts today—those who exhibit in the smartest galleries or whose dances, musical compositions, books, and plays are most celebrated, whose works command the highest prices, and those who write for the influential critical magazines—are for the most part united in their dissatisfaction with the "high-art" view. Loudly and clearly, they proclaim by deeds as well as words that hallowed notions of "fine art" and "aesthetic appreciation" are inadequate, misguided, parochial, chauvinistic, and even pernicious. . . . Still, despite the fresh air that postmodernist challenges have provided, reservations—not due simply to elitist snobbery or privileged Western chauvinism—remain. While one might concur broadly with their diagnosis that "high" or "fine" art repudiates or disregards much that is valuable in human imaginative and expressive activity, one must also agree that those who have found sustenance and even transcendence in their experience of the fine arts have good reasons to remain convinced that these sources and occasions are real and valuable. Repudiating the whole of Western civilization is a harsh response to admitted social inequities, so that postmodernist remedies that recommend flushing away the

baby with the bathwater can even seem more misguided than the malady *(2)*.

THE TRADITIONAL VIEW

In tracing the split between the traditional and the new views of art, the influence of reason stands out as a dominant factor. The old view of art is bound up with the notion of skill and craft. The practice of painting and medicine can each be described as art, and this idea of art being something skillfully done is implicit in words such as "artisan" and "artful." From these origins, over the course of the last millennium, the term "art" began to be applied more specifically to special skills and crafts which created something not intended for practical use—something with abstract qualities, such as beauty, that has a special value because it imparts a special feeling.

Art communicates feelings, but the feelings can undergo change in the course of transfer. Different people have different responses to the same work of art, and this erratic footing has given art the reputation of being at the mercy of the quirks of personal preferences and the foibles of arbitrary judgments. Such a conclusion discourages any rational analysis of art, but the premises of the argument are open to question, and people have engaged reason to solve the dilemma of subjectivity. In his essay on the criticism of art, I. C. Jarvie presents a case for objective value in art:

> To be a subjectivist and write criticism seems inconsistent. What would be the point of swapping opinions; and moreover opinions that are couched in language that suggests that they are more than opinions and nearer the truth? Possible answers are that the critic, thinking very highly of himself, believes his subjecttive opinions are worth more than those of others. But that is only his subjective opinion ... or he may be a complete cynic and do criticism merely as a job which he's grateful to have. None of these is very satisfactory *(7)*.

As an alternative, Jarvie argues that art can be looked at objectively, but he concedes that art is less rigorous than science: "An evaluation [of art] can be conclusively shown to be false (e.g., if it is self-contradictory) but it cannot be conclusively established as true, any more than can a scientific theory. Arguments can be deployed, but within limits; the regulative idea of truth is perhaps harder to catch and to know it has been caught than in science" (7). Artistic objectivity is not to be found in the opinion of individuals because everyone has personal biases; nevertheless, some kind of objectivity comes from the maturing effect of time, expressed through the establishment of artistic traditions and the formation of artistic institutions. The artistic traditions and institutions must be open to free criticism so that evaluations cannot be manipulated by external forces such as political or religious ideologies. Art must stand the test of time, so its judgment is necessarily a slow process. Over the years, however, a consensus is reached. According to this traditional view, we can be confident about old masters but we can never be sure about contemporary art.

THE NEW VIEW

The new view also uses reason, but it interprets the history of art differently. It sees all forms of art as having a direct social purpose. Cave paintings had religious significance as images designed to influence the success of a hunt. Prehistoric figurines were designed to improve fertility.

Church paintings were intended to deepen Christian thought and devotion. The church commissioned exactly what it wanted for its purposes, and if the artist failed to deliver, his work was rejected. When Caravaggio painted St. Matthew as an elderly, poor, working man—which he was—the church found this image of a saint unacceptable. Caravaggio had to paint another picture, in keeping with the church's concept of what a saint should look like (8).

After the Renaissance, Western paintings began to meet another need. They became objects through which their owners sought to establish and reinforce prestige, because paintings were fashionable and highly prized by patrons who could indulge themselves with luxuries. Just like the church, wealthy patrons exercised major control over both the composition of pictures and the financial arrangements. In his book *Painting and Experience in Fifteenth-Century Italy,* Michael Baxandall cites a letter from Filippo Lippi to Giovanni di Cosimo de'Medici, dated July 20, 1457:

> I have done what you told me on the painting, and applied myself scrupulously to each thing. The figure of St. Michael is now so near finishing that, since his armour is to be of silver and gold and his other garments too, I have been to see Bartolomeo Martelli; he said he would speak with Francesco Cantansanti about the gold and what you want. . . . I have had fourteen florins from you, and I wrote to you that my expenses would come to thirty florins, and it comes to that much because the picture is rich in its ornament. . . . to give me sixty florins to include materials, gold, gilding and painting, with Bartolomeo acting as my guarantor. . . . I send a drawing of how the triptych is made of wood, and with its height and breadth. Out of friendship to you I do not want to take more than the labour costs of 100 florins for this: I ask no more. I beg you to reply, because I am languishing here and want to leave Florence when I am finished. If I presume too much in writing to you, forgive me. I shall always do what you want in every respect, great and small *(9)*.

If this letter had been written recently, it could have come from an interior decorator—albeit one with exceptional courtesy. Alternatively, the alarming increase in cost is more suggestive of a plumber. Either way, the letter reveals the subtle border between craft and art, and how the commercial spirit was never far from the mind of either the painter or the purchaser. Baxandall catches this historical facet of art nicely: "paintings are among other things fossils of economic life" *(9)*.

———

MUSIC ALSO STARTED with a social purpose—to reinforce magic and religion. It was used to cast spells:

> Pliny reported that Cato had preserved an incantation for the cure of sprains, and Varro another for gout. Caelius Aurelianus mentioned the use of music in the general treatment of insanity and locally in the management of sciatica; "A certain piper would play his instrument over the affected parts and these would begin to throb and palpitate, banishing the pain and bringing relief." Caelius was sceptical and quoted an alternative view, "anyone who believes a severe disease can be banished by music and song is the victim of a silly delusion" *(10)*.

Gradually, like painting, music became a status symbol for the affluent and a way for them to celebrate their success. Again money was very important, and the patron could influence the work. Here one music publisher writes to another:

> By a little management, and without committing myself, I have at last made a complete conquest of that *haughty beauty* Beethoven . . . agreed with him to take in MS. three Quartets, a Symphony [the Fourth], an Overture [*Coriolan*], a Concerto for the Violin, which is beautiful, and which, at my request, he will adapt for the pianoforte, with and without additional keys; and a Concerto for the Pianoforte, for *all* of which we are to pay him two hundred pounds sterling.*

Literature, like painting and music, had its origins in religion—the Hindu Vedas, the Bible, and the Koran. Later, literature became the written record of the language prized by the elite. How is ordinary

---

* Letter (1807) of Muzio Clemente, a London publisher, to his colleague F. W. Collard, quoted by Jaques Attali, *Noise* (Minneapolis: University of Minnesota Press, 1996).

language turned into the special language of literature? Terry Eagleton illustrates the change:

> Literature transforms and intensifies ordinary language, deviates systematically from everyday speech. If you approach me at a bus stop and murmur "Thou still unravished bride of quietness," then I am instantly aware that I am in the presence of the literary. I know this because the texture, rhythm and resonance of your words are in excess of their abstractable meaning—or, as the linguists might more technically put it, there is a disproportion between the signifiers and the signifieds. Your language calls attention to itself, flaunts its material being, as statements like "Don't you know the drivers are on strike?" do not *(11)*.

Eagleton argues that "Thou still unravished bride of quietness" reveals the presence of a literary figure at the bus stop, but this is only true because he and the literary figure share the same background—they are of the same subculture. Others at the bus stop, confronted by this literary figure, might simply conclude that they are standing near a mental hospital.

THE ROOTS OF Western art grew out of a background of religion and then social exclusivity—a section of society reinforcing its status. In the past as in the present, people must be exposed to the appropriate subculture to enjoy its art—just as some level of literary awareness is required to appreciate the significance of being a "still unravished bride of quietness."

Although the appreciation of art depends upon shared cultural experience, which may sometimes be limited to an intellectually privileged group of cognoscenti, the new view sees the total domain of art as more extensive than previously. If the essence of art is the communication of feelings, then there is an element of art in religious rituals, political rallies, commercial advertisements, popular song, and popular fiction. This conclusion is in keeping with the egalitarian argument that no art form or style is inherently superior to any other, because art is fundamentally subjective. In contrast, the traditional view, presented by Jarvie *(7)*, holds that artistic traditions and institutions eventually establish a stable consensus on the merit of a work of art *(7)*. But the facts do not support Jarvie. There is no agreement on how to judge the quality of art. For want of anything better, prices give some indication of how a society values its art, but the prices assigned to classical, time-honored art, such as Renaissance paintings, have moved up and down dramatically—for example in the French auction rooms over the nineteenth century.

Here the new view is using reason very effectively by giving precedence to observations (the fluctuating price of art) over theory (traditional concepts of beauty). Proponents of the new view can extend their rational attack on traditional ideas by looking at the dominance of Western art. Where is the evidence that the sophisticated art of the West is better than the art of India, China, Japan, Bali, or the Queen Charlotte Islands? Each is different and each is specialized for its own culture. Each kind of art works best, in conveying emotions, when it is presented to people from its own culture. Thus the new view uses reason to question the existence of an all-embracing Art with a capital *A*. In contrast, it recognizes and champions particular forms of art—tied to particular cultures, in particular places—at particular times. It emphasizes the irrefutable fact that a villager from New Guinea will not experience the same emotions as a connoisseur of Western art when they walk together past French Impressionist paintings in the Metropolitan Museum of Art in New York. This is not surprising, but while the traditional view would regard the connoisseur's opinion as superior, the new view gives the opinion of the New Guinea villager equal standing, and would simply point to the reversal of roles if they continue their tour of the museum to visit the Michael C. Rockefeller collection of New Guinea art. We are not confined to the art of our own culture, but we should have some understanding of all the cultures whose art we wish to experience. The new view is epitomized by the trend for university departments of fine arts to change their names to departments of "visual cultures."

THE ARGUMENTS THAT underlie the new views of art are rational and persuasive. There seems to be a fundamental subjectivity in art that separates it from science. For unlike art, science stands demonstrably independent from its social and institutional embodiments—however much the purveyors of antiscience would like to deny this. Science works because there are objective laws that govern the objective facts of the world; some of these laws can be discovered, and some of them can be applied. In contrast, it is subjective feelings rather than objective facts that speak for art—just as feelings rather than facts speak for religion. The more we compare art and religion, the more features we find in common. Religion and art vary from culture to culture, yet each has its own universal features present in all cultures. Both religion and art specialize in creating enhanced states of mind that may sometimes achieve ecstatic intensity. Both have temples (museums and galleries) where people congregate and participate. Both have high priests (critics) who interpret and who command authority. Both have bigotted zealots, and both have more numerous humble devotees.

WHY IS ART IMPORTANT?

Compared with the other human endeavors that we have been discussing—morality, commerce, government, and religion—art seems to be less essential. Although we can agree that society needs art, some might consider it a luxury to be indulged for recreation. But if it is a luxury, it is a very old one, going back 40,000 years. Art takes time—for training and execution. Our ancestors must have put most of their energy into activities urgently necessary for survival, so why did they divert time and resources for art? Art must have contributed to evolutionary success, but how? We do not have a definite answer, but we can guess. In early human communities art must have been

so closely interwoven with religion that it would have been difficult
to tease the two apart. Prehistoric figurines were art designed to increase fertility; prehistoric wall paintings were art designed to bring
success in hunting. Prehistoric music and dance were art designed to
reinforce religious rituals, and rituals "endow culturally important
cosmological conceptions and values with persuasive emotive force,
thus unifying individual participants into a genuine community"
(12). Art and religion were inseparable, so the biological role of art
was the same as the biological role of religion—to strengthen cultural bonding. The role remains similar today, for although art may
have developed a life of its own separate from religion; still it remains
a bonding force that holds cultures and subcultures together.

If we accept the new view, the cultural reinforcement achieved by
classical art emerges as a small, important detail in a much larger picture. But now we can broaden our perspective and accept as art all
those activities that bring individuals together to share emotional experiences—other than religion. We can include popular song and
popular fiction under the umbrella of art, so many more people become involved. Furthermore, spectator sports have much in common with the performing arts and they should be added to the
conglomerate, which now becomes a highly influential human activity. The biological purpose of this expanded concept of "big art," and
its value in conferring evolutionary advantages, stem from its ability
to reinforce social cohesion. When we looked at how the diverse elements of British society came together at the Battle of the Somme,
we saw enormous cultural strength built by unifying an assortment
of subcultures for a common purpose. We can speculate that this
kind of group solidarity was even more important over the hundreds
of thousands of years of our preliterate history.

11

SCIENCE

There is a real world independent of our senses;
the laws of nature were not invented by man, but forced upon him
by that natural world. They are the expression
of a natural world order.

—MAX PLANCK*

SCIENTIFIC PRINCIPLES UNDERPINNED ancient engineering feats such as the construction of the great Egyptian tombs and temples. The Greeks formalized the rational basis of science, and then the focus of activity moved back to the Middle East. Independently and simultaneously, science and technology were rising in China.

With the Renaissance and the Enlightenment, science returned to Europe and the growth of knowledge about the nature of the world accelerated. It is easy to underestimate the difficulties encountered by early scientists. Herbert Butterfield provides perspective:

---

* Max Planck was awarded the Nobel Prize for physics in 1918. He stayed in Germany during the Nazi era, without giving any support to the regime. His younger son, Erwin, was arrested by the Gestapo and executed for alleged complicity in the 1944 plot to kill Hitler. This quotation is attributed to Planck by Max Perutz, another Nobel laureate, in his book *Is Science Necessary?* (London: Barrie and Jenkins, 1989).

The greatest obstacle to the understanding of the history of science is our inability to unload our minds of modern views about the nature of the universe. We look back a few centuries and we see men with brains much more powerful than ours—men who stand out as giants in the intellectual history of the world—and sometimes they look foolish if we only superficially observe them, for they were unaware of some of the most elementary scientific principles that we nowadays learn at school *(1)*.

How did science evolve? Emile Durkheim viewed the growth in the size and complexity of human populations as the driving force. Early, small communities had to concentrate all their physical and mental effort on survival; their thoughts were focused on food and religion. As communities became larger, some people had time to reflect and debate. They found they could understand and predict events better if they reduced passion and prejudice, replacing these with observation and inference. But while a large population may have been necessary, in itself it was not sufficient for science to germinate. The Roman and Chinese empires were big, but the rigid social control required to hold an empire together was not conducive to science, just as it was not conducive to reason. The early nurturing and later flowering of science required a large *and* loosely structured, competitive community to support original thought and freewheeling incentive. The rise in commerce and the decline of authoritarian religion allowed science to follow reason in seventeenth-century Europe.

Widespread acceptance of science as an independent system of thought is relatively modern. In the words of Bertrand Russell written in 1951:

Science, as a dominant factor in determining the beliefs of educated men, has existed for about 300 years; as a source of economic technique, it has existed for about 150 years. In this brief period it has proved itself an incredibly powerful revolutionary force. When we consider how recently it has risen to power, we find ourselves forced to believe that we are only at the very beginning of its work in transforming human life *(3)*.

The church was probably the strongest pillar sustaining medieval European culture, and it repressed early science, which it regarded as a threat. Slowly, the church began to realize that science was not a fashion that could be stamped out because it offended traditional ideas. The predictions made by science usually proved to be right—this was bound to give it special status and influence. Furthermore, scientific predictions were not of a trifling nature. The patrons of science reaped its rewards in commerce and warfare so science became indispensable. As the church saw what was happening, its attitude gradually changed. Cultures all over the world followed the trend set in Europe, because they had no option if they wanted to compete on the international stage. Science became accepted as a normal and essential part of life.

Francis Bacon's aphorism *"Nam et ipsa scientia potestas est,"* proclaimed in 1597, is usually translated as "Knowledge itself is power," but as we embark on the twenty-first century, we could recast it as "Science itself is power" *(4)*. Science is now an industry of such vital importance that every country has to assign a significant portion of its budget to educating young people in its ways. Science has become a great locomotive that can take us to certain destinations, but it cannot select these destinations. Our cultures set the route and make up the timetable. Some destinations are not accessible at present—it is no good asking science to cure cancer if we do not have a track in the right direction. The fact that a routing is unclear does not mean that science has failed; it merely means the terrain has not been adequately explored and there are many obstacles to be cleared.

OVER TIME, THE very success of science paradoxically became a problem, for it generated unrealistic hopes that inevitably led to disappointment and the growth of an antiscience movement. In order to understand what we should and should not expect of science, we must look further into its nature. If philosophy and mathematics are the most direct *theoretical* expressions of reason, then science is the most direct *practical* expression of reason.*

Thomas Huxley† encapsulated the essence of science as: "nothing but trained and organized common sense, differing from the latter only as a veteran may differ from a raw recruit; and its methods differ from those of common sense only in as far as the guardsman's cut and thrust differ from the manner in which a savage wields his club" *(5)*. Huxley was more specific in another metaphor: "The chess-board is the world; the pieces are the phenomena of the universe; the rules of the game are what we call the laws of nature. The player on the other side is hidden from us. We know that his play is always fair, just and patient. But also we know, to our cost, that he never overlooks a mistake, or makes the smallest allowance for ignorance" *(6)*. Scientists apply themselves to analyzing the rules and playing the game, but it often takes them into unexpected and unknown territory. This gives science an exploratory dimension, one discovery leading to another without orderly direction. "Modern science has imposed on humanity the necessity for wandering . . . from generation to generation, a true migration into uncharted seas of adventure" *(7)*.

Science, then, is organized common sense, an analysis of the laws of nature, and a voyage of exploration. It is also, as Einstein and Planck both insisted, a commitment to the existence of an external

---

* As Gary Gutting put it: "The structures of reason are apparent not in abstract principles but in the concrete employments of reason. Norms of rationality are constituted in the very process of applying our thoughts to particular problems, and science has been the primary locus of success in such applications." See G. Gutting, *Michel Foucault's Archaeology of Scientific Reason* (Cambridge: Cambridge University Press, 1989).

† Thomas Huxley was the son of a London schoolmaster. The school failed, and the family moved to Coventry where his father struggled with a new job in a Savings Bank. Thomas started to study medicine, and received the gold medal in the first part of the examination for his medical degree, but he did not have enough money to continue his medical education, so he turned his attention to biology. He became a staunch supporter of Charles Darwin, and he debated evolution with Bishop Wilberforce at a meeting of the British Association in Oxford in 1860. See Adrian Desmond, *Huxley: The Devil's Disciple* (London: Michael Joseph, 1994).

world, independent of the perceiving human subject. The workhorse of science is inductive reasoning. In a traditionally scientific setting the formal argument runs: "When I ignited a mixture of hydrogen and oxygen yesterday it formed water; therefore hydrogen ignited with oxygen forms water; so if I ignite hydrogen with oxygen tomorrow it will form water." The structure of this argument is not in keeping with the real world unless the role of chance is acknowledged: "When I ignited a mixture of hydrogen and oxygen yesterday it *usually* formed water; therefore hydrogen ignited with oxygen *usually* forms water; so if I ignite hydrogen with oxygen tomorrow it will *usually* form water." The element of chance can be expressed with precision through the application of probability theory and statistics. Chance is a challenge to science: "The characteristic of phenomena that we call fortuitous, or due to chance, is that they depend upon causes that are too complex to enable us to know and study them all" *(8)*. Probability theory underlies statistics, and it also underlies Heisenberg's uncertainty principle—that it is impossible to determine, exactly, both the position and the momentum of a particle at any instant in time. The more accurately we measure position, the less precise is the estimate of momentum, and vice versa, so the product of the two errors is a constant.

While argument plays a central role in the operation of science, observation always takes precedence—Galileo's knife. Francis Bacon made the point nicely: "Argumentation cannot suffice for the discovery of new work, since the subtlety of nature is greater than the subtlety of argument"* *(9)*.

---

* The preeminence of observation in the development of knowledge is profound yet frequently ignored. For example, medicine is still linked to magic by faith healers. Mrs. C.J. of Dayton, Ohio, sent the following letter to Senator Matthew Neely when he promoted a prize for the cure of cancer: "Dear Sir, In reading the paper I saw a reward for the cure of cancer. Not that I am after the money, but just to show you what the Lord will do, I am sending an anointed handkerchief, and if you do as I tell you, you will be cured of the disease. Now, just lay it over the cancer in the name of Jesus, and it is healed if you believe it. If this doesn't do any good, it is because you have no faith." Quoted in S. P. Strickland, *Politics, Science, and Dread Disease* (Cambridge, Mass.: Harvard University Press, 1972).

The student of science learns that its method starts with observation and ends with observation, with important stages of imagination and argument in between. After an initial observation, the scientist must engage original thought—imagination—to construct a hypothesis. Experiments are designed to test the hypothesis. Since no experiment can prove anything, emphasis is placed on attempting to disprove the hypothesis—refutation of ideas about the real world is always possible while certainty can never be attained. Karl Popper has underscored this principle, insisting "that not the *verifiability* but the *falsifiability* of a system is to be taken as a criterion of demarcation. . . . *(10)*. So experiments are performed and observations are made. If the working hypothesis withstands all attempts to refute it, new knowledge can be claimed. This is a stepping stone toward a better understanding of the world, and toward new technology. In this way we have gained new concepts, new laws about nature, and new disciplines such as quantum mechanics and molecular biology. The practical benefits are obvious—motor cars, computers, and increased supplies of food and energy. Medical science has produced vaccines, antibiotics, anesthetics, and X rays. A long list of infective scourges have come under control—poliomyelitis, smallpox, diphtheria, typhoid, tuberculosis, leprosy, puerperal fever, yellow fever, and malaria.

THE ELEMENTS OF reason are the infrastructure of science—induction, deduction, Galileo's knife, and Ockham's razor. The superstructure is built from extensions of reason—an imaginative constituent to creat hypotheses, the method of experimentally attempting falsification, and the application of probability theory. These are, collectively, scientific universals, just as we have language universals and ethical universals. Science is not a culture specific enterprise, for there are no fundamental differences between American science, British science, Russian science, or Chinese science, beyond the obvious fact that the scientists happen to work in different places.

International scientific meetings do not concern themselves with *where* science is done—they discuss *what* science is done.

NEW KNOWLEDGE OF the world is subject to modification by future scientists. As pointed out by Einstein and Infeld:

> Creating a new theory is not like destroying an old barn and erecting a skyscraper in its place. It is rather like climbing a mountain, gaining new and wider views, discovering unexpected connections between our starting point and its rich environment. But the point from which we started out still exists and can be seen, although it appears smaller and forms a tiny part of our broad view gained by the mastery of the obstacles on our adventurous way up *(11)*.

With this background, we shall look at the operation of science, taking an example from neurological research. Equipped with the microscope, tissue stains, electrical stimulators, and surgical scalpels, nineteenth-century scientists explored the structure and function of the brain. Equipped with microelectrodes, biochemistry, immuno-cytochemistry, and molecular biology, twentieth-century scientists have continued to give us a marvelous understanding of how the brain is organized. Applied science has revolutionized the treatment of psychiatric illnesses such as schizophrenia and depression, and neurological disorders such as Parkinson's disease and epilepsy. Sometimes the advances have come from a carefully planned series of studies, but the astute recognition of a lucky observation has been equally important. We shall now look at both these methods, taking the planned approach first.

The discovery of levodopa, as a treatment for Parkinson's disease, illustrates how a rationally organized coordinated series of observations can lead to a useful conclusion. (1) In the late 1950s dopamine was identified as a neurotransmitter (a chemical agent used to transfer information from one nerve cell to another); (2) soon after, post-mortem examination of brains from patients with Parkinson's disease revealed a selective loss of dopamine; (3) a "blood-brain barrier" impeded the passage of many substances (including dopamine) to brain cells, so treatment had to cross this barrier and restore the concentration of dopamine; (4) the ideal agent proved to be the substance that nerve cells normally use to make dopamine—levodopa; (5) tests on patients then showed that levodopa was highly effective in relieving the symptoms of Parkinson's disease. Each step of the story sounds simple, but each was a challenge. For example, the first therapeutic tests on patients began in 1961, but it was not until 1967 that the real benefit of levodopa emerged; the formal credentials required for accepting new treatment—efficacy and safety demonstrated in a double-blind clinical trial—took another two years.

How were the methods of science applied in the development of levodopa? Let us examine the first step, the discovery that dopamine is a neurotransmitter. Dopamine was known to be produced in the course of making other neurotransmitters in the brain, so its presence was originally thought to reflect the synthesis of these agents. Then came the observation that dopamine was concentrated in distinct regions of the brain, separate from the areas rich in the other neurotransmitters. Hermann Blaschko grasped the significance of this finding and suggested that perhaps dopamine had "some regulating functions of its own"—in other words it might be a neurotransmitter in its own right (12). Experiments were designed to test this hypothesis, and nobody was able to disprove it. The scientific community therefore accepted, as new knowledge, the conclusion that dopamine is a neurotransmitter, and the stage was set for the pivotal discovery, by Oleh Hornykiewicz and his colleagues, that dopamine is depleted in the brains of patients with Parkinson's disease (13).

Of course, science does not always proceed as smoothly as this. All too often new ideas are refuted—as Thomas Huxley put it, "the great tragedy of Science—the slaying of a beautiful hypothesis by an ugly fact" *(14)*. Peter Medawar said the same in his *Advice to a Young Scientist:* "There is often no way of telling in advance if the day-dreams of a lifetime dedicated to the pursuit of truth will carry a novice through the frustration of seeing experiments fail and of making the dismaying discovery that some of one's favorite ideas are groundless" *(15)*. There are many blind alleys and many mistakes; observations may be inaccurate and arguments may be flawed. The story of levodopa therapy is unusual because it proceeded without any major setback. More often things go wrong, so tenacity is the key to success.

As in all fields of human endeavor, luck can make or break an enterprise. Many important observations have been made by chance. Each such case can be broken down into the following three steps: (1) chance presented a particular observation; (2) a general conclusion was drawn; (3) a predicted outcome was tested. For example, patients with severe Parkinson's disease cannot swallow their saliva normally so they tend to drool. In 1867 Jean-Martin Charcot, a great French neurologist, tried to treat this drooling with hyoscyamine, a product of deadly nightshade that reduces the secretion of saliva. L. Ordenstein, a student, reported what happened to the patients' tremor: "From the beginning of this month, Monsieur Charcot prescribed one or two granules of hyoscyamine, each amounting to about 1 mg. This medication produced some hours of respite [from tremor] for many patients" *(16)*. Quite unexpectedly, Charcot had found a treatment for the tremor of Parkinson's disease, and similar treatment remained the cornerstone of anti-Parkinson therapy for a century.

The most recent development in the treatment for Parkinson's disease is an interesting hybrid of luck and induction. For hundreds of years European communities had been struck down by poisoning from rye infected with a fungus, *Claviceps purpura*. Patients complained of burning pain in their fingers and toes, so the sickness was called St. Anthony's fire. Over the present century pharmacologists identified the ergot alkaloids as the toxic substances in the fungus. Plant toxins—like the hyoscyamine just discussed—often have medical applications when given in low doses, so the actions of the ergot alkaloids were analyzed. The results were encouraging: the ergots caused uterine contraction that proved useful in treating hemorrhage after childbirth. Ergot derivatives were also helpful in migraine. A third property was suppression of lactation. These three actions seemed puzzlingly unrelated, so pharmacologists began to separate them and analyze each mechanism. The task was undertaken in stages, and the study of lactation yielded results that were to prove unexpectedly important for patients with Parkinson's disease.

The first challenge was to discover how the ergots worked; it turned out that they decreased the release of prolactin, a hormone that promotes lactation. Once the mechanism of action was known, a series of candidate chemicals was made and tested to find one with optimal properties—most effective, least toxic, and cheap. The result was a compound called bromocriptine. Further studies showed that the secretion of prolactin was normally controlled by a "prolactin inhibitory factor" acting in the pituitary gland. The next step was the discovery that the prolactin inhibitory factor was actually dopamine—a finding that proved to be the decisive "chance" observation suddenly linking all this work to Parkinson's disease. Once the prolactin inhibitory factor was known to be dopamine, it became clear that bromocriptine achieved its effect on lactation by stimulating cells that respond to dopamine in the pituitary gland. Perhaps bromocriptine would stimulate other brain cells that respond to dopamine. Corrodi and colleagues therefore tested the effect of bromocriptine on an animal model of Parkinson's disease (17). The results were encouraging, so bromocriptine was given to patients. It is now a widely used anti-Parkinson's drug.

Because levodopa and bromocriptine have the same pattern of efficacy but a different profile of side effects, optimal treatment for most patients is a combination of both drugs in low dosage—rather than a high dose of one. Current treatment of Parkinson's disease thus has its origins in two strategies. First, in the case of levodopa, we saw the rational orchestration of a focused scientific effort whose direction was clear from the start. Second, in the cases of hyoscyamine and bromocriptine, we saw how unexpected observations created opportunities that were exploited in a rational way.

IN BROADER PERSPECTIVE, Einstein and Infeld have likened science to a great unsolved mystery story:

> The reading has already given us much; it has taught us the rudiments of the language of nature; it has enabled us to understand many of the clues, and has been a source of joy and excitement in the oftentimes painful advances of science. But we realize that in spite of all the volumes read and understood we are still far from a complete solution, if, indeed, such a thing exists at all. At every stage we try to find an explanation consistent with the clues already discovered. Tentatively accepted theories have explained many of the facts, but no general solution compatible with all the known clues has yet been evolved. Very often a seemingly perfect theory has proved inadequate in the light of further reading. New facts appear, contradicting the theory or unexplained by it. The more we read, the more fully do we appreciate the perfect construction of the book, even though a complete solution seems to recede as we advance *(11)*.

Science has grown to such an extent that it is no longer a single coherent body of knowledge. Hilary Putnam has portrayed it in a naval metaphor, with researchers at sea in a fleet of boats.

> The people in each boat are trying to reconstruct their own boat without modifying it so much at any time that the boat sinks. . . . In addition, people are passing supplies and tools from one boat to another and shouting advice and encouragement (or discouragement) to each other. Finally, people sometimes decide they do

not like the boat they are in and move to a different boat altogether. And sometimes a boat sinks or is abandoned. It is all a bit chaotic; but since it is a fleet, no one is totally out of signaling distance *(18)*.

The boats are of different size and different stability, and they move at different speeds, but the image is not entirely chaotic, for they all steer the same course forward into the unknown. Enterprises classed as "big science," the aircraft carriers of Putnam's fleet, are expensive, prominent, and difficult to operate efficiently because of their size. They have a high profile that makes them vulnerable, so they draw most of the fire from the enemies of science.

External critics of science tend to be destructive, but criticism from within is usually constructive. Karl Popper draws attention to the benefits of this self-criticism: "The history of science, like the history of all human ideas, is a history of irresponsible dreams, of obstinacy, and of error. But science is one of the very few human activities—perhaps the only one—in which errors are systematically criticized and fairly often, in time, corrected" *(19)*.

These various views of science illustrate what science is and what it does. One important question remains—if the practice of science is fraught with frustration and disappointment, as it certainly is, why do people make it their lives' work? Einstein offers an answer:

In the temple of science are many mansions, and various indeed are they that dwell therein and the motives that have led them thither. Many take to science out of a joyful sense of superior intellectual power; science is their own special sport to which they look for vivid experience and the satisfaction of ambition; many others are to be found in the temple who have offered the products of their brains on this altar for purely utilitarian purposes. Were an angel of the Lord to come and drive all the people belonging to these two categories out of the temple, the assemblage would be seriously depleted, but there would still be some men, of both present and past times, left inside. . . . A finely tempered nature longs to escape from personal life into the world of objective perception and thought; this desire may be compared to the townsman's irresistible longing to escape from his noisy, cramped surroundings into the silence of high mountains, where the eye ranges freely through the still, pure air and fondly traces out the restful contours apparently built for eternity *(20)*.

DIFFICULTIES WITH SCIENCE

Of course there are limits on what kind of question science can answer, and failure to appreciate these limits has led to disappointment and misplaced criticism. Detractors of science also point accusing fingers at ecological damage, nerve gas and atomic bombs. But science cannot control the way it is used. It is a tool for reaching goals, not for setting them. Like reason, science is a servant that will attempt to do whatever is asked. When governments became concerned about pollution, science created catalytic convertors and commercial chimney scrubbers. Cultures have to decide whether they want science to work on central heating or flamethrowers. Science is neither moral nor immoral—it is impartial.

The critics of science are not satisfied with this, for antiscience has deeper roots. (*21*). Recently a mathematical physicist, Alan Sokal, made a reconnaissance sortie into the territory of the antiscientists. He published a paper in the journal *Social Text* entitled: "Transgressing the Boundaries—Toward a Transformative Hermeneutics of Quantum Gravity (*22*). The paper is a hoax and a parody of postmodernist antiscience, yet it was accepted for publication because, in the view of its author, "(a) it sounded good and (b) it flattered the editors' ideological preconceptions" (*23*).

Sometimes good ideas of one scientist are not understood or accepted by others. This may slow down scientific developments, but only temporarily. Edward Jenner's first paper on vaccination was rejected by the Royal Society in 1797. David Horrobin cites further examples of intellectual rigidity among scientists:

> In 1879 the Marquis of Sautuola, an extremely distinguished amateur archeologist, discovered the spectacular cave paintings of Altamira. He privately circulated news of his find and was promptly labeled a fraud and a forger. He was not only not allowed to present his findings at international congresses of archeology, but he was actually banned from attending them and died in 1888, a disappointed man. Only in 1903 did any colleague decide to check the story and demonstrate that the paintings were indeed by early humans and not by a modern forger (24).

Similar difficulties persist:

1. The article by Glick et al. (25) on the identification of B lymphocytes as separate entities is one of the seminal papers in immunology. It was rejected by leading general and specialist journals and eventually appeared in *Poultry Science* because of the species in which the work was done. *Poultry Science* is a respectable and respected journal, but perhaps not the place where one would expect to find such a fundamental article.

2. Krebs' article (26) on the citric acid cycle, possibly the most important single article in modern biochemistry, was initially rejected by the peer review process.

3. The work of S. A. Berson, M.D., and Yalow (27) on radioimmunoassay, which, like Krebs' studies, eventually led to a Nobel Prize, was initially rejected for publication (24).

In 1993, Michael Smith received the Nobel Prize for chemistry. This was awarded for his discovery of a way to create mutations at selected sites along a strand of DNA. His original observations were made in 1978, and when he sent his findings for publication in the prestigious journal *Cell*, his manuscript was rejected with the comment "a technical development of no general interest" (personal communication).

These examples of foundering science are disconcerting but they do not mean that science is inherently flawed; they simply show that scientists, like the rest of us, can become too set in their ways. From time to time scientists also lie, and, as in any walk of life, there are a few cheats. These human failings are inevitable in all human enterprises; they do not detract from the methods or the achievements of science any more than forgers detract from the methods or the achievements of art.

12

BEHAVIOR AND THE BRAIN

The ruling passion, be it what it will,
The ruling passion conquers reason still.
—ALEXANDER POPE*

C AN ANIMALS REASON? Or, more precisely, are the adjustable patterns of animal behavior the building blocks of rational thought? Research on our closest living relatives, chimpanzees, gives us the best insight on the intellectual capacity of our ancestors. Premack and Woodruff made videotapes of humans attempting to solve problems such as trying to reach bananas or get out of a cage. The videotapes were shown to chimpanzees, who were then presented with photographs that included solutions to the problems. The chimpanzees consistently chose the pictures that provided the answers *(1)*. This study suggests that chimpanzees can reason (they may, incidentally, be wondering whether people have the capacity for reason, and if so, why they need help from apes to solve such simple problems).

We can see the broad shape of reason in more familiar animals. Dogs are adept at recognizing changes in their owner's actions that signify a walk may be in the offing and they immediately try to hasten it. They search for whatever is most likely to have the desired ef-

---

* "To Lord Bathurst" (1733), line 155, *Epistles to Several Persons.*

fect—a leash, a hat, a glove, or a boot. This is versatile goal-directed behavior; surely it can be interpreted as a larval form of reasoning?

Instinct is at the opposite end of the spectrum to reason; it is the least versatile kind of behavior, and it is the oldest we can find in evolution. Somewhere and somehow reason arose from this background, for complex biological systems cannot just appear without leaving a trail in the record of evolution. Reason is the latest adaptation that animals have found useful to help them adjust to a changing environment.

INSTINCT

From the time of conception, we are each endowed with our own complex, powerful, and indelible array of genes—a blueprint inherited through millions of years of evolution by our ancestors—and the blueprint includes instinctive drives. Genes are molecules that form templates for the manufacture of proteins, and proteins establish the structure and control the function of cells. As cells grow, they multiply and interact to form tissues. Nerve cells develop in the fetus and establish contact to form enormously complex networks. Genetically organized groups of nerve cells are activated for such purposes as seeking food, finding a mate, and rearing the young. Such instinctive behavior is based upon a stereotyped pattern characteristic for each species.

The pattern, however, can be modified by experience with the environment. Examples abound of behavior that is the outcome of genetic drives interacting with environmental experiences. In humans, particularly, the distinction between the influence of nature and nurture is often blurred. This flexible concept of instinct contrasts with the old classical idea of a fixed stimulus leading to a fixed response. The demise of the rigid notion of instinct took place gradually over the last eighty years, and the concept of instinct being hard-wired in the brain was almost abandoned. In 1919 K. Dunlap wrote a paper entitled "Are There Any Instincts?" (2) and in 1955 F. A. Beach entitled his review of the subject "The Descent of Instinct" (3).

Recent advances in molecular biology have reversed this trend, because the power of genetic control is now recognized in so many fields of science, including neuroanatomy, neurophysiology, psychology, psychiatry, and neurology. A new image of instinct has emerged.

A. N. EPSTEIN defines three components. First, there are internal factors, such as nutrients and hormones circulating in the blood, and the genes defining the patterns of needs and responses for the species. Second, a reactive system enables an external stimulus to trigger the onset of a response. Third, the individual's emotional state, personal history, and cultural background may modify the expression of an instinct.

From this analysis we can see that a stimulus by itself is not enough to elicit an instinctive reaction—the stimulus is necessary but not sufficient. Internal factors, namely a state of responsiveness and a genetic framework for the response, are also necessary but not sufficient. The external stimulus and internal factors are together sufficient, but the final result is under the sway of emotions, cultural attitudes, and experience.

If, for one reason or another, instinctive drives cannot be satisfied, animals usually pursue different behavior—a phenomenon called *displacement.* In humans the equivalent redirection is called *sublimation,* a term most commonly used when the sexual drive is frustrated in a way that results in extra energy going to activities such as work or sport. Psychoanalysts refer to this kind of reaction as a defense mechanism; they have developed theoretical constructs in which anxiety follows failure to redirect pent-up instinctive pressure.

In biological terminology, "depletion-repletion" portrays the oscillating physiological status that underlies instinctive behavior—thirst and hunger are typical examples. From this background a model for more complex human instinct designates "drive" as the internal state and "incentive" as the external stimulus. Because instinct is the behavior that we can find so abundantly in the animal kingdom, it must bear some ancestral evolutionary relationship to reason. We shall, therefore, pause to look at some of the features of specific instincts, starting with self-preservation.

THE FORMULA FOR survival includes food, drink, a suitable environmental temperature, and the ability to elude predators. *Homo sapiens* has all these needs, and for most of humanity over most of history, getting food must have been the highest priority. We can catch a glimpse of what it is like to be in chronic need of food by seeing how current impoverished communities exist on the edge of starvation. Robert Thouless recounts the way the Sironi, a group of seminomadic Bolivian Indians, deal with hunger as the dominant factor in their lives:

> Food and food getting are reported to be the subjects of their major anxieties. When food is obtained, it is eaten furtively and without ritual. Conflicts about food are the greatest single cause of conflict within the group. Food and successful hunting dominate their dreams and phantasies, while sex dreams and phantasies rarely occur. They have little sexual activity while food is scarce, but sexual orgies follow a successful hunt. They show sparse development of art, folk tales and mythology; their magic principally relates to food (5).

While the psychology of hunger is complex, the physiology of hunger is not much simpler taking into account the automatic adjustment of appetite necessary to sustain an optimal body weight for survival. The key mechanism for controlling hunger is, appropriately enough, a feedback loop. Chemical changes in the blood influence specialized receptors in the brain to set in motion a pattern of exploratory activity, such as hunting or gathering, that culminates in feeding. Food is then digested and absorbed into the blood. The resulting change in blood constituents stimulates the brain to switch off food-seeking behavior—although a few species have an instinctive drive to continue collecting food which they put aside for hard times.

The physiology of thirst involves sensors that recognize a reduction in the volume of circulating blood and dehydration of tissues. At a certain level the brain initiates behavior that is directed to finding water. In unusually dry climates evolution has gone beyond the instinct for seeking water; an entire ecosystem has evolved so that when it rains in the desert an assortment of animals and plants spring to activity from beneath the sand, and when the rain stops, life resumes its low profile.

Animals can react instinctively to predators—when danger is recognized, a program for "fight or flight" is abruptly set in motion. In some species, automatic attack is triggered if the other animal is retreating, automatic retreat if the other animal is advancing. The individual needs self-preservation above all else, but the species has different priorities—the genes must be passed on *(6)*. When reproductive behavior conflicts with self-preservation, reproduction usually prevails. For example, bright coloring helps insects to find each other for mating, although it also attracts insect eaters. Similarly, parents risk their lives to ensure the survival of their offspring—the future of the species depends upon the young more than the old. Of course this principle cannot be taken to extremes, for many species the young survive better if the old are there to look after them.

In keeping with the central importance of reproduction, courtship is a significant feature of animal life. A classic example is the mating behavior of the three-spined stickleback. The male fish becomes extremely possessive over its territory in the mating season. It guards a portion of the stream and attacks any intruding male. Experiments with models of fish show that aggression is triggered by the sight of the red belly of another male stickleback. In contrast, the appearance of a female stickleback elicits a complex, stereotyped zigzag courting dance.

ABRAHAM MASLOW HAS suggested that instinctive needs normally establish themselves in a hierarchy, and a glance at human behavior gives support to this idea *(7)*. In broad brush strokes, hunger, thirst, and reproduction are fundamental. Only when these are satisfied will the next category, the need for security, emerge. When we have security, social needs arise, such as belonging, loving, and being respected. Finally, there are drives to explore and to achieve. All needs are under the sway of cultural influences, which define their precise expression. For example, what constitutes "achieving" will differ between New Yorkers in Manhattan and Dayaks in Borneo. Cultural forces can also manipulate the hierarchy of instincts, so some people, for example, will sacrifice everything to climb Mount Everest.

Yet for most of us there is a limit to how far cultural ambitions can push the biological hierarchy of motivation. In the last few years sexual scandals have rocked the reputation of princes, ministers, and presidents. Modern bloodhounds are armed with telescopic lenses and DNA tests, and they are paid generously by intrusive media and politicized investigators. This technology is known by the public figures who are targeted, but in spite of the risk of exposure, the instinctive urges are too strong to be denied. In wider perspective, there is abundant evidence of the triumph of sexual desires over cultural restraints. From divorce courts and medical records we must conclude that although social conventions try to play down its influence, sexuality is one of the most powerful forces in our lives. Before the era of antibiotics, syphilis was a scourge afflicting the rich and poor of all races and creeds. People were told "For one pleasure, a thousand pains"—they knew the danger yet they pressed on. The impotence of reason in this setting, and the ability of sexual drives to prevail over all cultural taboos, is a vestige of our biological past, when vigorous reproduction was crucial for the survival of every previous ancestral species in the history of our evolution.

INSTINCTS BECAME ESTABLISHED by natural selection; their variety reflects how nature automatically selects the drives most useful for survival. The discovery of genetic mutation reveals how inherited diversity arises. Richard Dawkins has suggested that the genes assert self-interest ruthlessly, using the bodies of animals as machines to carry them, increase their numbers, and pass them on *(6)*. Our bodies may last eighty years, but our genes can endure for millions. While all this is true, the genes are not parasites because they make a vital contribution to the life of the organism. The genes need the organism and the organism needs the genes—there is mutual interdependence. The overwhelming pressure to go forth and multiply is equally beneficial for the species and the genes.

The process of evolution is one of continually adapting to a changing environment. There is no progress—if this means advancing toward a goal—because there is no goal other than survival. Maynard Smith explains what happens, through his concept of the "evolutionarily stable strategy" *(8)*. New circumstances lead to instability within a population of animals, and in response the population undergoes change. When an optimal adjustment is achieved, stability is reestablished.

EMOTION AND MOOD

The essential difference between emotion and reason is that emotion leads to actions while reason leads to conclusions. The distinction between reason and emotion is so steeped in tradition that for centuries reason was believed to reside in the brain and emotion in the heart. But now we know that the heart is just a pump. The emotions share their home in the brain with reason.

The power of emotions has been recognized since the beginning of recorded history. Emotions and moods are closely related but differ in duration; emotions are transient and moods are prolonged—emotions are like the weather, while moods are like the seasons. Emotions, like instincts, crave satisfaction. They move us to seek or avoid what may loosely be termed psychological rewards or punishments—things that make us feel good or bad. This central feature of all emotions separates them from reason, which has no craving, no drive, and no demand for satisfaction.

What, exactly, are emotions? It is easier to give illustrations than to provide a definition. Paul Ekman has analyzed facial expressions and from the broad spectrum of patterns he has suggested that there are six universal or "basic" emotions: happiness, surprise, fear, sadness, anger, and disgust *(9)*. More complex emotions can be assembled from these building blocks, just as three primary colors can be mixed to create any color of the rainbow. A list of composite emotions includes affection, anxiety, boredom, disappointment, envy, guilt, hate, indignation, irritation, love, panic, shame, and sorrow.* Such forces have propelled our enterprises throughout history.

In COLIN MCGINN's framework of mental activities, emotions fall into the category of "sensations" *(10)*. Certainly, emotions are closely associated with alterations in bodily feelings. William James went so far as to say that if we abstract from our consciousness all the bodily feelings associated with emotions, there is nothing left *(11)*. De Sousa also emphasized the importance of bodily sensation in emotional experience:

Ordinary talk about emotions, from slang to poetry, is replete with physiological allusions. Consider, for example, the following common expressions, all of which are used to describe emotion: "What a pisser," "to shit a brick," "to wet your pants," "my throat went dry," "my heart was pounding," "to break out in a cold sweat," "to blush, or flush," "to get hot around the collar," "my blood ran cold." . . . Sometimes the correlation between organs and emotions is more literary than literal: "in my day, my dear, the organ of love used to be the heart." But consider the following description:

---

* These emotions crave reassurance (anxiety, panic, disappointment, indignation, guilt, shame, sorrow), reciprocation (affection, love), hostility (envy, hate) or excitement (boredom).

When I see you my voice fails
my tongue is paralyzed,
a fiery fever runs through my whole body
my eyes are swimming,
and can see nothing
my ears are filled with a throbbing din
I am shivering all over . . .

Is this a protocol from a psychophysiological laboratory? No. It is
a literal translation of some lines about lust by the poet Sappho,
written some twenty-six centuries ago *(12)*.

Emotions, however, are more than simple sensations of bodily
functions. Like instincts, emotions are engines of motivation. They
make us take action because they entail mental rewards (feelings of
satisfaction)—or mental punishments (feelings of frustration). We
are moved to increase feelings of satisfaction and to decrease feelings
of frustration. Stated in such starkly simple terms, motivation seems
trivial, but it is profoundly important. The emotions drive passionate love, religious ecstasy, and the laying down of one's life for one's
country. Yet the urge to satisfy oneself is not always virtuous, for it
also drives greed, crime, and, of course, drug abuse.

In addition, emotions provide a rapid system of communication:
they let other people know what we feel and what we may do. James
drew attention to the ease with which the outward manifestations of
emotion can be recognized and he also argued that instincts merge
into emotions:

In speaking of the instincts it has been impossible to keep them
separate from the emotional excitements which go with them.
Objects of rage, love, fear, etc., not only prompt a man to outward deeds, but provoke characteristic alterations in his attitude
and visage, and affect his breathing, circulation, and other or-

ganic functions in specific ways. When the outward deeds are inhibited, these latter emotional expressions still remain, and we read the anger in the face, though the blow may not be struck, and the fear betrays itself in voice and color, though one may suppress all other signs. *Instinctive reactions and emotional expressions thus shade imperceptibly into each other. Every object that excites an instinct excites an emotion as well* (11; emphasis in original).

Normally, emotions are so transparent that we are continuously and automatically adjusting our behavior to take account of other people's feelings. We do not laugh if the faces around us are sad; we do not hesitate to laugh when the faces around us are happy. Emotions can cross the boundary of species. We readily identify joy, fear, or anger in dogs, and dogs seem to have no difficulty in identifying the same emotions in us.

THE VALUE OF emotional responses, as tools for survival, is illustrated by the cliché that they prepare us for fight or flight. But how do we decide between these alternatives? Our instinct for self-preservation provides the answer. "Fight" if you can win and "flight" if you can't. While some automatic behavior is directed by our genes, other automatic reactions can be acquired. As Damasio points out:

Consider what happens when we move away briskly from a falling object. There is a situation that calls for prompt action (e.g., falling object); there are options for action (to duck or not) and each has a different consequence. However, in order to select the response, we use neither conscious (explicit) knowledge nor a conscious reasoning strategy. The requisite knowledge was once conscious, when we first learned that falling objects may hurt us and that avoiding them or stopping them is better than being hit. But experience with such scenarios when we grew up made our brains solidly pair the provoking stimulus with the most advantageous response. The "strategy" for response selection now consists of activating the strong link between stimulus and response, such that the implementation of the response comes *automatically* and *rapidly,* without effort or deliberation *(13)*.

Sometimes, however, the circumstances do not give a clear signal for immediate action—the encounter may yield inadequate or confusing information. Here, prior cultural exposure may elicit an emotional response, or reason may be recruited to assess the situation. D. Goleman has compared the "emotional mind" with the "rational mind":

The emotional mind is far quicker than the rational mind, springing into action without pausing even a moment to consider what it is doing. Its quickness precludes the deliberate, analytic reflection that is the hallmark of the thinking mind. In evolution this quickness most likely revolved around that most basic decision, what to pay attention to, and, once vigilant while, say, confronting another animal, making split-second decisions like, Do I eat this, or does it eat me? Those organisms that had to reflect too long on these answers were unlikely to have many progeny to pass on their slower-acting genes *(14)*.

Emotions, like instincts, exercise a stern authority, yet emotions are readily manipulated. We all have an emotional response of anger, but what makes us angry is influenced by how and where we are brought up. In recent history, the capacity for a culture to manipulate an emotion is illustrated by the anti-Semitism of the Third Reich. By skillfully orchestrating a propaganda campaign it proved possible to promulgate a hodgepodge of truths, half-truths, and untruths so that the Jews became a national scapegoat and the universal stimulus for hatred. With their machinery of indoctrination, the Nazis were able to pass laws and execute policies of such malevolence that they now seem scarcely credible. Yet within one generation of the fall of Hitler a new German culture recast the image of the Jews to elicit the emotion of guilt.

Information on brain activity underlying the emotions is limited. Robert Heath has recently recorded microsynchronization of neuronal activity in people experiencing rage. This microsynchronization occurs in regions of the brain called the amygdala and the hippocampus *(16)*. Twenty years earlier Heath had found similar activity during human orgasms, in a deep central portion of the brain—the septal area *(17)*. We have already discussed microsynchronization in the temporal lobe as a possible physiological basis for ecstatic religious experiences, and we have speculated that similar activity may underlie intense aesthetic experiences.

LEARNING

As young children learn to interpret their senses they are explorers embarking on a voyage of discovery. They orient themselves within a framework established by their concepts of space, time, and causation. They learn to speak, and they gain the ability to interpret the behavior of others. Children are repeatedly performing experiments in statics (seeing how high things can be piled on top of each other), dynamics (seeing how far things can be pulled, pushed, or dropped), ballistics (seeing how far things can be thrown), and psychology (pushing parental responses to the limit). Children learn more rapidly than adults as they enthusiastically study the ways of the world.

Before exploring learning in more detail, we should pause to make a distinction between behavior that is determined by selecting from a repertoire of genetically established options, and behavior that is acquired through experience. It now seems likely that many of our mental and motor responses are genetic, or in the jargon of the cognitive scientists "Domain specific Darwinian algorithms."* More patterns of thought and behavior may be selected from the genetic stock pile than we previously imagined, but learning from experience still extends the range and versatility of our options.

The traditional concept of learning entails building up associations from experience to achieve *new* patterns of behavior and *new* thoughts.† The capacity for collective learning and storage in the memory of a culture is particularly powerful, because it allows knowledge to accumulate so quickly.

Learning is a process of profiting from experience. Perhaps the most important step in the evolution of learning was the emergence of pleasure and pain as mental experiences *(18)*. Pleasure is linked to events that contribute to survival, while pain is linked to events that are harmful. By associating pleasure with survival and pain with extinction, the scene was set for an expansion of the biological capacity for profiting from experience. Alexander Bain suggested that pleasure and pain become associated not only with the stimulus but also with the response, and even with the "idea" of what had taken place *(19)*. Reexperiencing any one of the components—the stimulus, the response, or the mental image of the event—rekindles pleasure or pain. How are these linkages established—between pleasure, pain, the environmental situation, the action taken, and the conse-

---

* These are specialized brain circuits evolved by evolutionary selection pressures. Michael Gazzaniga argues by analogy to immunology, where we know that many antibody molecules are coded by our genes, so the presentation of an antigen new to the individual leads to the selection of the appropriate antibody from the immense range available in our "reference library." Our DNA holds the evolutionary wisdom accumulated over millions of years, just as our culture holds the environmental wisdom accumulated over thousands of years. We can select from either source as necessary. See M. Gazzaniga, *Nature's Mind* (New York: Basic Books, 1992).

† Recently psychologists have started to emphasize that complicated behavior can also be achieved by computing information. For example, ants and bees perform navigation by repeatedly checking their speed and direction from the time they leave their nests, rather than learning their way through experience. See Steven Pinker, *How the Mind Works* (New York: Norton, 1997).

quence? In short, how does the process of learning take place? There seem to be several ways, with very different potentials and levels of complexity.

Habituation. In terms of evolution, the oldest and simplest form of learning is habituation—the ability to develop a modified response when the same stimulus is repeatedly presented in the same way. Habituation occurs in very primitive animals, even those without a nervous system. It plays a small part in the behavior of human subjects, but it can still be demonstrated quite easily. For example, if a startling stimulus is repeated several times, the response steadily declines.

Imprinting. Another primitive form of learning was first studied by D. A. Spalding in the last century.* In their first few days of life, chickens follow any moving object, and then their behavior reverses so that they run away from anything that moves, other than the object that they started following initially. In normal circumstances, this means that the chicks follow their mother and avoid everything else, because their mother is the closest moving object after birth. The evolutionary value of this "imprinting" is obvious.

The Conditioned Reflex. In the seventeenth century Descartes promoted the idea that a stimulus can elicit a response through a mechanism analogous to reflection by a mirror—the reflex. He added that this phenomenon can be modified by the soul, acting through the brain. Subsequent studies confirmed the existence of reflexes and showed that many can still occur in animals that have been decapitated, so the spinal cord became the seat of reflex function. It was not until the nineteenth century that Ivan Sechenov proposed the exis-

---

* The observations were reported in D. A. Spalding in "Instinct—with original observations on young animals," *Macmillan's Magazine* 27 (1873): 282–293. Konrad Lorenz made further observations on the phenomenon eighty years later.

tence of two types of reflex:* (1) a relatively primitive and inflexible group of reflexes that primarily involve the spinal cord, but may extend into portions of the brain that are, in evolutionary terms, old—the paleocortex; and (2) a more complex and versatile group of reflexes that use newer regions of the brain—the neocortex—to generate responses that are modifiable by experience.

The stage was set for Pavlov to demonstrate *conditioning*.† First he showed that dogs normally salivate when stimulated by food. Then he identified a stimulus, the ringing of a bell, that was not linked to food or salivation. He repeatedly combined the two stimuli—the ring and the food—and found that soon the dogs would salivate when he rang the bell without offering the food. This process of experimentally creating new reflexes, by association of stimuli, is a simple form of learning.

Operant Conditioning. The conditioned reflex requires rather precise timing of two stimuli together, an unusual situation in normal life. A more complex and powerful type of learning is built by associating stimuli with *rewards* or *punishments* in a loose temporal relationship. In this way rats can be trained to press a bar and to obtain food, drink, or even a sexual partner. Human parents routinely administer rewards and punishments as they rear their children—the first time an infant defecates into a pot, parental approval

---

* Ivan Sechenov was the father of Russian physiology, but much of his important work was carried out in Claude Bernard's laboratory in Paris. His studies were reported in his book *Physiologische Studien über die Hemmungsmechanischen für die Reflexthätigkeit des Rückenmarks im Gehirne des Froches* (Berlin: A. Hirschwald, 1863).

† Ivan Pavlov was the son of a sexton. He received the Nobel Prize in 1904 for his work on digestive secretions. Although he lived all his life in Russia, he openly denounced communism. In 1922 he asked Lenin to let him transfer his laboratory to another country, but Lenin refused. Pavlov proclaimed, "For the kind of social experiment you are making, I would not sacrifice a frog's hind legs." He denied Bucharin, the Soviet commissar for education, permission to visit his laboratory, although the laboratory was funded by the government through Bucharin. Later Pavlov became less hostile to the regime. See Jeffrey A. Gray, *Ivan Pavlov* (New York: Viking Press, 1980).

leads to repeat performances. The same approach is adopted by teachers and psychologists in schools, clinics, and prisons. The term *operant conditioning* (or *instrumental conditioning*) has been coined for this way of learning. An important difference between a conditioned reflex and operant conditioning is that a stimulus initiates the former (the ringing bell), while the individual initiates the latter (the rat pressing a bar). In both cases encouragement or discouragement—positive or negative reinforcement—plays a vital role in the outcome.

Operant conditioning is a normal mechanism of learning, and the most stable, resilient behavior is established by intermittent, irregular, and unpredictable reinforcement. This is clearly advantageous from an evolutionary viewpoint because our environment is subject to frequent changes—the ability of a lion to learn from often, but not always, finding its prey at a watering hole is biologically useful; if it had to find its prey there every time in order to learn, it would go hungry.

Negative reinforcement, or punishment, is the traditional method of dealing with unwanted behavior. The standard treatment for addiction to alcohol is the administration of disulfiram (Antabuse). This drug interacts with the products of alcohol to cause nausea and vomiting. For someone who is receiving disulfiram, the results of drinking are unpleasant enough to achieve the required effect. Alcoholics who are unable to apply reason to their problem can nevertheless respond to a program of therapeutic punishment of this kind.

Modeling. Simple observation of events, particularly the actions of other individuals of the same species, can be enough to modify behavior without going through a process of conditioning. Totally new patterns of behavior can be learned by *modeling* in this way. Time and again children learn by modeling their behavior on adults, and adults also learn by copying other adults. The entire advertising industry is a giant exercise in modeling. Young and attractive individ-

uals are shown enjoying the right beer or cola. Old and respectable
couples are shown displaying their kindness and generosity by using
telephones to call their grandchildren. The message is loud and clear:
if we are to be attractive, happy, relaxed, satisfied, and successful, we
should follow the examples before us.*

THIS BRIEF SURVEY provides a glimpse of how brain mechanisms'
underlying behavior may be broken down into parts. Nature (predisposition from our genes) and nurture (experience from our environment) combine to determine our behavior, just as they combine
to determine the shape of our bodies. We can see how instinct, emotion, and learning give us an infinite number of ways to react. Out of
this background reason emerged, and its power grew as the brains of
our ancestors enlarged.

THE ORIGIN OF THE HUMAN BRAIN

How did the human brain evolve? First, we must look at our position
in the classification of the animal kingdom. We are in the order *Primates,* the family *Hominids,* the genus *Homo,* and the species *sapiens.*
The family Hominidae comprises the genera *Australopithecus*
("southern apeman," of which none survive) and *Homo* (of which
the only surviving species is *Homo sapiens*). In tracing the origin of
the human brain, we will start by creating perspective in time. The
Roman emperor Marcus Aurelius put it well: "Every instant of time
is a pinprick of eternity. All things are petty, easily changed, vanish-

---

* While we may recognize the profit motive behind advertisements, we are less quick to
realize that much of our entertainment, unrelated to advertising, is still geared to titillate
salacious thrills. Are these harmful to ourselves or society? There is no compelling evidence, but people still hold strong views, on each side of the vigorous debate over watching sex and violence on television. Tannis MacBeth has edited a recent series of opinions,
and she summarizes them in a cautiously worded introduction: "Many different groups
of experts, brought together over the years, have reached the conclusion that although no
single study can be definitive, the body of research evidence indicates that television does
have some effects on some viewers under some circumstances." See T. M. MacBeth, ed.,
Tuning In to Young Viewers (Thousand Oaks, Calif.: Sage, 1996).

ing away."* The age of the universe and the course of biological evolution are measured by scales so different from our normal idea of
time that they are scarcely comprehensible. Fossils show that animal
life began some 600 million years ago. Each of the recent geological
periods lasted less than 100 million years and when a period ended,
more than 99 percent of the existing species died out. The less than
1 percent of survivors generated a range of new species that diversified and outstripped the previous number, leading to an increase in
biological variation. The vertebrates started about 500 million years
ago, with hominids dating back 4 million years.

The brain began as a swelling at the front of the spinal cord. The
convoluted superficial layer, the cerebral cortex, was the most recent
addition. Over the course of mammalian evolution, the cortex has
expanded in area a thousandfold, with little increase in thickness. In
general, increasing body mass correlates with the possession of larger
brains; but if we plot the size of brain against body weight, there are
two outlying groups—primates and dolphins have big brains.

The brain receives information from the outside world via the
eyes, ears, nose, mouth, and touch receptors. Appropriate responses
to external events arise from a neurophysiological process that ultimately generates patterns of impulses which are sent down nerves to
muscles or glands. Control of the right side of the body comes, predominantly, from the left side of the brain, and vice versa. Most of
the impulse traffic to and from the brain must, therefore, change
sides. This crossing of the major pathways is, of course, of great importance to neurologists and neurosurgeons. It means that when the
right side of the body becomes paralyzed, the cause must be sought
in the left side of the brain. Comparative anatomical studies indicate
that crossing has very early evolutionary origins, going back through
mammals, birds, fish, and reptiles to animals that existed before the

* Marcus Aurelius was Roman emperor from A.D. 161 to 180. This quotation is from his
Meditations, a rare record of a Roman emperor's thoughts. Marcus Aurelius studied law
and philosophy, and he was unusually generous. He was one of the best emperors that
Rome produced.

development of backbones. Crossing gave our aquatic ancestors a mechanism for rapid escape from predators. Detection of a threat on one side of the body led, through crossed pathways, to contraction of the trunk muscles on the opposite side. This, in turn, caused the body to arch away from the threat.

Particular functions are localized in anatomically distinct portions of the brain. The cerebral cortex has highly specialized areas; microscopic examination reveals a consistent pattern of clustered nerve cells with specific tracts of nerve fibers running to, from, and through them. While many evolutionary influences on the brain can be recognized as changes in size or structure that are visible to the naked eye, such alterations represent only the tip of the iceberg. Beneath the microscope, evolution has been continuously remodeling the circuitry. Certain neural pathways, such as those contributing to manual dexterity, have been of great value to hominids, so natural selection has operated to preserve and enhance them. Other pathways, such as those responsible for high olfactory sensitivity, are less important so they have declined. A concept of "neural Darwinism" has been put forward by Gerald Edelman to explain how interconnected groups of nerve cells concerned with different functions have evolved in the cerebral cortex (20).

When did reason appear in this burgeoning growth of neural circuitry? Intelligence correlates roughly with the size of the brain, so we should now turn our attention to the animals with the largest brains for their body weight, the primates. The suborder Anthropoidea contains the most recently evolved primates: monkeys, apes, and humans. The monkeys are quadrupeds with tails; the apes are bipeds without tails. As we get closer to *Homo sapiens* in evolution, lineage becomes more difficult to establish. The distinction between species and race is particularly hard in early hominids, yet it is important. Different races can interbreed while different species cannot. It is not easy to decide when a species appeared and disappeared; fossil bones are one guide, but there are other relevant archeological findings, such as stone tools. Sometimes the tools extend beyond the period of the anatomical fossils. Does this mean we simply have not discovered enough fossils, or were the same tools used by more than one species?

Ethiopia has the earliest fossils of apeman and apewoman, *Australopithecus afarensis*. This species appeared 4 to 4.5 million years ago and lasted for 1 to 1.5 million years. The next hominid species was *Australopithecus africanus*, whose era ran from 3 to 1.5 million years ago. *Australopithecus africanus* blends with *Homo habilis*, the first member of our genus. *Homo habilis* existed from 2 to 1.6 million years ago, and had better control over the environment, achieved through a "tool-using brain"—an enlarged area of the cerebral cortex representing the hand. This step also required an increased range and precision of finger movements, so in addition to having a more powerful brain, *Homo habilis* had a more skillful hand with redesigned bones, joints, and muscles, particularly for the thumb. The evolutionary significance of manual dexterity was comparable to the importance of language. In the words of Thomas Carlyle, "Man is a tool-using animal. . . . Without tools he is nothing, with tools he is all" *(21)*.

Homo habilis was followed by *Homo erectus*, from 1.8 million to 300,000 years ago; recent studies suggest that the species may even have survived up to 27,000 years ago. *Homo erectus* was taller, more mobile, and more adventurous, exploring the world as the first hominid to undertake intercontinental travel on two feet, from Russia, China, Java, and Indonesia. *Homo sapiens* appeared some 400,000 years ago. For the first 200,000 years there were significant differences from the modern species, such as more prominent brow ridges; because of these distinctive features early humans are often called "archaic" *Homo sapiens.*

A century ago the fragments of a skull were found in the Neander valley near Dusseldorf. The skull was initially thought to belong to an early Celt, a victim of rickets, an "idiot," or a Cossack. Even the great anatomist Rudolf Virchow failed to recognize the antiquity of the find; he dismissed the skull as a deformed modern human. Eventually Neanderthals were recognized as a separate group, but their story remains a puzzle.* Fossilized evidence suggests that the Neanderthals emerged about 150,000 years ago, and they became extinct some 30,000 years ago. Their brains were about the same size as ours, but their eyebrows and jaws were more prominent. Their limb bones were stronger and more suitable for traveling long distances over rough terrain.

Neanderthals were concentrated in Europe and seem to have thrived in cool climates. They left behind a great variety of high-quality stone tools. We do not know the origin of Neanderthals, their relationship to *Homo sapiens,* or the reasons for their disappearance. One suggestion attributes their extinction to difficulty in dealing with warm climates. We know that *Homo sapiens* spread to the areas inhabited by Neanderthals so we cannot exclude another possibility—that violent clashes resulted in genocide. People living in close proximity readily become hostile to each other, particularly if they have a different physical appearance. The same forces that underlie modern racism are likely to have erupted when *Homo sapiens* came face to face with Neanderthals—"they look different from us; they behave differently from us; they compete with us; they threaten to undermine our way of life."

THE SIZE OF the brain increased dramatically from *Australopithecus afarensis* to *Homo sapiens.* The australopithecan brain was 500 milliliters in volume, comparable to that of chimpanzees and gorillas. *Homo habilis* had a brain of about 600 milliliters, *Homo erectus* about 900 milliliters, and *Homo sapiens* 1,200–1,700 milliliters. Over a period of 4 million years the size of the hominid brain increased threefold—an astonishing rate of change. The factors leading to hominid

---

* At present there is a controversy among anthropologists over whether Neanderthals were a distinct species. The disagreement illustrates the difficulty of drawing firm conclusions from bones and stones.

survival were intellectual development, manual dexterity, and language. The rapid increase in size and complexity of our ancestors' brains was attained without weakening its structure; it remained rugged and functional throughout life. Isaac Watts's metaphor strikes an appropriate chord: "Strange! that a harp of a thousand strings should keep in tune so long."*

While the human brain is larger than the brain of chimpanzees, there is no evidence to suggest that *Homo sapiens* has any different type of nerve cell or any different neurochemical constituent. The human nervous system differs from that of apes only in number and organization of nerve cells. Enlargement of the hominid brain was not a generalized process engaging all regions equally. Growth in the size and complexity of the front of the brain outstripped any other region—the frontal areas connect different parts of the brain with each other, so the competition for survival favored more "brain power" rather than better vision, better hearing, or a better sense of smell.

Because of the increasing size of the brain, later hominids were born with larger heads, and this caused difficulties. There was a limit to how wide the birth canal could become, since enlarging the pelvis made walking and running awkward. Children had to be born at an earlier stage of development, which, in turn, meant more careful and prolonged rearing, so the family structure had to be more stable.

Savanna grasslands dominated the East African landscape as hominids began to separate from apes. In this environment, natural selection favored an upright posture and long legs, to increase mobility and field of vision. The presence of big game in the grasslands gave a biological advantage to hominids who could organize themselves into powerful hunting groups—powerful because they had better tools and better weapons and communication. Further-

---

* Isaac Watts (1674–1748), an English preacher, is remembered for the hymns he wrote, which include "O God Our Help in Ages Past."

more, their improved intelligence enabled them to plan and adjust the hunt in ways that had never before been possible.

WE DO NOT know the extent to which the various hominids developed in parallel or in sequence. Evolutionary "branches" can form new "twigs" so that closely related species exist at the same time, possibly competing with each other. Recent studies of DNA on the Y chromosome suggest that all present humanity came from a small community in Africa; racial diversity emerged only within the last 100,000 years (22). No primary human race has survived; the Australian aborigines probably resemble early *Homo sapiens* more closely than any other race. The three major races, Caucasoid, Mongoloid, and Negroid, are all composite; they arose through interbreeding within communities that were isolated from each other for tens of thousands of years.

Some of the differences between the races may have arisen by chance, such as the high cheek bones in Mongoloids. Others, such as pigmentation of the skin in Negroids, are likely to have arisen as biological adaptations. In hot climates there is a high level of exposure to sunlight, carrying a significant risk of skin cancer. This danger is increased by the high temperature because people wear few clothes. Skin cells containing the brown pigment, melanin, offer protection, so there is a survival advantage in having dark skin. In contrast, cold climates have less sunlight and people cover their bodies with clothes, but here another problem arises. *Homo sapiens* needs to absorb sunlight through the skin to produce vitamin D. Lack of vitamin D leads to rickets, so it is better to have less skin melanin—hence the advantage of pale skin.

Intelligence is an advantage in all human communities, and all races have a similar share. Detailed comparison of intelligence among races is fraught with methodological problems because of the difficulty in finding a test that is not influenced by different cultural backgrounds. In the present discussion there is nothing to be gained from entering the bitter, highly politicized debate over whether one or another race has a small lead or handicap in the various rating scales used to measure intelligence. All investigators agree that there are no major differences.

The racial divisions and subdivisions merge into each other. Just as Darwin identified natural selection as the driving force of evolution, he also saw the variation of racial diversity: "It may be doubted whether any character can be named which is distinctive of race and is constant" *(23)*. The wisdom of this opinion is even more evident now that more intercontinental travel and less cultural prejudice leads to more mixing of the gene pool.

THE TRADITIONAL IMAGE of our early preliterate history presents biologically evolved *Homo sapiens* adopting a social way of life and building a human culture. Clifford Geertz has summarized this view:

> Man's physical being evolved, through the usual mechanisms of genetic variation and natural selection, up to the point where his anatomical structure had arrived at more or less the status at which we find it today: then cultural development got under way. At some particular stage in this phylogenetic history, a marginal genetic change of some sort rendered him capable of producing and carrying culture and thenceforth his form of adaptive response to environmental pressures was almost exclusively cultural rather than genetic *(24)*.

Geertz goes on to criticize this interpretation and replace it with an alternative more in keeping with the view presented here, that a primitive culture existed as long ago as the time of the Australopithecines, and rather than being superimposed on *Homo sapiens,* culture contributed to the creation of the species. This was achieved through the kind of feedback loop we have already discussed, whereby modules of brain structure for language, manual dexterity, and reason were linked to the benefits they brought. An initial, marginally effective language module in the brain led to improved social

interaction, which, in turn, led through natural selection to an improved version of the language module in the brain, and so on. This
sort of interaction could explain the exceptionally rapid expansion
of the brain in the four million years between *Australopithecus
afarensis* and *Homo sapiens*.

STRUCTURE AND FUNCTION OF THE BRAIN

Most of the ancient Greeks, including Aristotle, believed that the
brain secreted phlegm and cooled the blood. But Hippocrates had
greater insight, giving the brain responsibility for intelligence,
thought, and dreams; in short, he placed the mind firmly in the
brain: "Men ought to know that from the brain, and from the brain
only, arise our pleasures, joys, laughter, and jests as well as our sor-
rows, pain, griefs and tears. Through it in particular, we think, see,
hear and distinguish the ugly from the good, the pleasant from the
unpleasant."* Thomas Willis restated and extended this view: "The
Brain is accounted the chief seat of the Rational Soul in a man . . . and
indeed as the chief mover in the animal machine, it is the origin and
founder of all motions and conceptions."†

Santiago Ramón y Cajal surveyed the brain under the microscope;
he was the master of mapping neuroanatomy. He depicted the brain
as "a world consisting of a number of explored continents and great

---

* Hippocrates, famous for his "oath," lived on the island of Cos around 400 B.C. Some
seventy treatises were attributed to him; they were written by his followers over a period
of about 200 years. This quotation comes from his treatise "The Sacred Disease"
(epilepsy), one of his best-known works. The translation is by W. H. Jones, *Hippocrates*,
vol. 2 (London: Heinemann, 1923).

† Willis published his classical anatomical description of the brain, spinal cord, and
nerves in *Cerebri Anatome* (1664). It was translated into English in 1681. Willis coined the
term "Neurologie" in this treatise: "To describe all the several pairs of the spinal Nerves,
and to rehearse all their branchings, and to unfold the uses and actions of them, would
be a work of an immense labour and trouble: and as this *Neurologie* cannot be learned
nor understood without an exact knowledge of the Muscles, we may justly here forebear
entering upon its particular institution." Willis is also credited with being the first to report the clinical features of temporal lobe epilepsy.

stretches of unknown."* He established the nerve cell as the structural unit of the brain. The body of the cell has numerous branches that receive several thousand contacts from other nerve cells. One specialized portion, the nerve fiber, arises from the cell body and extends over distances ranging from a millimeter to a meter—depending upon where it is situated—before it branches into multiple terminals. Each of these endings establishes close contact with another nerve cell. Electrical nerve impulses are unable to jump across this junction between nerve cells; instead, the nerve endings release a chemical agent, called a neurotransmitter, that crosses the gap. This transmitter reacts with receptors on the surface of the adjacent nerve cell.

For each transmitter there are several types of receptor, each produced by its own gene. For dopamine there are five types; for another transmitter, glutamate, twenty-one different types have been identified. Some receptors respond to the presence of a neurotransmitter, such as dopamine, by initiating events to promote a new impulse in the next nerve cell, while others have the opposite effect—in this way nerve cells can communicate "excitation" or "inhibition." Each nerve fiber releases one major type of transmitter and some may also release secondary agents that modify the action of the primary transmitter. The nerve fibers run in bundles along circumscribed pathways. Many unrelated pathways use the same ensemble of transmitters, so the 10 billion nerve cells in the brain can operate efficiently with a total of less than 50 chemically distinct neurotransmitters—a highly efficient arrangement. As a further economy, some of the same chemical agents that communicate information at the junctions between cells, such as dopamine, are also employed as hormones in the endocrine system.

Thus the biological machinery is in place for one nerve cell to con-

---

* This quotation comes from F. H. Garrison's translation of *Chácharas de Café*, originally published in 1923. Ramón y Cajal was born in northern Spain. His father was a barber-surgeon; his childhood was notable for his mischief. He received the Nobel Prize in 1906 for his atlas of the microscopic anatomy of the brain.

tribute to switching the next cell "on" or "off," as in the binary tran-
sistor circuit of a computer. But communication between nerve cells
is not confined to switching "on" or "off"; the quantity of neurotransmitter delivered by one impulse at one nerve terminal is so
small that its influence depends upon how many other packages of
transmitter are released by other impulses in the same microscopic
region at the same instant in time. This arrangement allows communication between nerve cells that is graded into minute steps. Although the nerve cells in the brain have far more extensive and
flexible functional connections than any computer, there is still some
similarity in the way brains and computers operate. The technology
of microchips has brought us the concepts of modular organization
and parallel processing; the brain uses both of these arrangements
for handling large quantities of information and executing complicated tasks.

Modular organization means that the brain has specialized components made up of neural circuits that deal with different categories
of information independently. For example, language is executed by
a module that is specialized to construct and interpret sentences.
Naturally, the output from the various modules must be bound together so that the brain ultimately functions as a whole.

Parallel processing means that related tasks are undertaken simultaneously in different parts of the brain. For example, the coordination of voluntary movement is achieved through specialized
activities in the cerebral cortex, the basal ganglia, and the cerebellum.
These regions collectively feed patterns of nerve impulses into centers that integrate the converging information and channel it into
"final common pathways" to the muscles.

FORMATION OF MODULES

We know little about how the developing brain creates modules.
Nerve cells multiply vigorously in the early fetus, so there is a stage
when the brain has many more cells than it needs; redundant components then die away. If they persisted, the excess of nerve cells

would disrupt normal function. So the developing brain is a micro-cosm of natural selection. Many nerve fibers grow toward specific targets—other nerve cells with which they must make contact. For each pathway, the fibers that grow establish connections, but when all the available points of contact in the target area have been taken, nerve fibers arriving late have nowhere to go. In this microscopic race for survival, the losers—the nerve cells that do not reach their target in time—disappear without trace.

Several mechanisms seem to be controlling events as the brain de-velops. At an early stage, specialized chemical agents promote adhe-sion so that nerve cells clump together, and other substances, called trophic factors, promote the survival of nerve cells, their growth, and the sprouting of nerve endings. Sometimes nerve endings growing along different pathways intermingle and interact—a phenomenon called "handshake." Time is crucial for establishing the architecture of the brain; certain stages of gestation and postnatal development offer a window of opportunity for specific spurts of growth.

The wiring diagram of the brain evolves as construction proceeds. First, there is a genetic blueprint that defines the broad brush strokes of the design through a process that is independent of the activity of nerves. Second, there is the fine brushwork that is equally essential but depends entirely on the presence or absence of nerve activity. Cells that fire together, wire together, and pathways are readily re-shaped if necessary. In biological terms, there is a high level of plas-ticity. This plasticity allows major readjustments—the *structure* of the brain is changed by its *function* at critical stages of development. For example, the anatomy of the brain has been shown to change if baby animals are socially isolated.* The classical early experiments on plasticity were performed on the visual pathways. Many of the

---

* L. J. Martin and colleagues studied the brains of monkeys raised in social isolation over the first nine months of life. They found a reduction of certain nerve cells. See "Social De-privation of Infant Rhesus Monkeys Alters Chemoarchitecture of the Brain," *Journal of Neuroscience* 11 (1991): 3344–3358.

cortical nerve cells concerned with vision respond to stimulation of either eye. If one eye is covered temporarily, abnormal neural connections are established so that after removal of the cover, cortical cells fail to respond to stimulation of this eye. In contrast, nerve cells responding to the side that was not covered display enhanced response.*

Genetic programming and neuronal activity combine to produce a series of modules. The modules are sufficiently complete in themselves for their functions to be selectively impaired by neurological disease. But each module also links up with others, so the entire brain is a complex modular organization of neuronal circuitry. Patrick Doherty and colleagues give a simple summary of the process of building a brain: "Whereas the coarse hard-wiring of the nervous system appears to rely on molecular recognition events between the neuron, its pathway, and its target, the establishment of precisely patterned functional circuits is thought to be driven by neuronal activity" (25). When there is contact with the outside world after birth, stimulation increases activity in functionally related nerve cells so that the structural integrity of modules is assured. The "use it or lose it" principle, at specific stages in development, underlies "the most complicated wiring task in the world" (26).

Nerve cells lose the ability to replicate before the end of pregnancy, and after making contact with each other, they must settle down for their long-term responsibilities: the business of transmitting nerve impulses and looking after themselves. The brain consists of billions of nerve cells, and there are thousands of billions of connections, but each cell has its own needs. It must generate energy from nutrients in the blood and it must dispose of waste products. It must also repair damage from wear and tear.

---

* In 1981 David Hubel and Torsten Weisel received the Nobel Prize for this discovery. Their definitive report was "Single-cell Responses in Striate Cortex of Kittens Deprived of Vision in One Eye," *Journal of Neurophysiology* 26 (1963): 1003–1017.

INTEGRATION OF BRAIN ACTIVITY

If we seek a general impression of how the brain operates, we find an intricately assembled organ undertaking a multitude of immensely complicated tasks simultaneously. Charles Scott Sherrington described it as "an enchanted loom where millions of flashing shuttles weave a dissolving pattern, always a meaningful pattern though never an abiding one" *(27)*. Sherrington's enchanted loom conjures up an image of extraordinary delicacy and fragile balance, yet we know that most of the nerve cells of the "enchanted loom" usually sustain their structure and function through some three score years and ten. A more prosaic metaphor for the brain can be assembled from modern computer jargon. We have a central processing unit in our heads, with hardware circuitry made up from nerve fibers and nerve cells. The input is from sensory stimuli that tell us what is happening in the environment. The output is mental activity, physical activity, and the control of bodily functions. The operating systems are provided by our genes. Our experiences are stored in data banks of memory. Reason is a program for data analysis; it is innate, in the same way that the program for operating language is innate. Reason and language have a hard-wired module of brain structure, whose designs are to be found in the genetic blueprint of our species. Reason is not inherited in a final form, any more than people inherit the English, French or Italian language; it is only the operating systems for reason and language, their underlying mechanisms, that are inborn.

To illustrate the way the brain integrates its activities, we can examine how it deals with a simple routine event, a daily "good morning" greeting so trivial that we scarcely notice it has taken place. While walking on the street, my eyes record a pattern of light reflected from a face. A coded sequence of nerve impulses is carried from my retinae to my brain. The intensity of light reaching the retina is not important, for I can recognize a face equally well in the sunshine or the shade. It is the pattern of activity among the retinal cells that is important. The corresponding pattern of nerve impulses in the brain is linked to memories of faces that I have seen before, and processed in the context of previous experience—the time, place, and circumstance in which my memories of faces are set. I recognize the face as that of a friend. The resulting activity in my brain is combined with information drawn from my capacity for grammar, my store of words, and my recall of the conventions of behavior normally associated with meeting someone I know. I assemble a pattern of nerve impulses that is sent to the muscles of my larynx, pharynx, and diaphragm, and I say, "Good morning."

While all of this is taking place, my brain continues to perform other essential housekeeping activities. The movements of my eyes are being coordinated to maintain stereoscopic vision; complex and continuous modification of activity in my body musculature is taking place to adjust my posture and balance; and my rate of breathing is being controlled to ensure that the levels of oxygen and carbon dioxide are appropriate. These and countless other automatic tasks are under way, each requiring information to be processed so that suitable instructions can be sent for appropriate actions.

Another category of an automatic event taking place is the assignment of priorities to different kinds of information, to protect the processing system from overload. In the unremarkable social event just described, the encounter started when my eyes sensed light reflected from a face, but there was also light reaching my retinae from all manner of additional objects—sky, buildings, motor cars, and innumerable other potential distractions. The brain engaged a filtering process so that precedence was given to a privileged minority of components out of the vast array of incoming information. If there had been a road traffic accident in the background, this would have become the focus of attention instead of my friend's face. Automatic monitoring and evaluation of my perceptual screens would have reformulated the priorities and my behavior would have been entirely different.

The mechanisms depicted here, underlying a routine morning greeting, are, of course, grossly oversimplified. Perception is complex and highly subjective; we see (1) what is biologically important, (2) what we look for, and (3) what we have been trained to see. Even such details as the transfer of nerve impulses from the eye to the brain are oversimplified in this description of seeing a friend. There is not *one* region of the cerebral cortex that receives the incoming impulse traffic; there are *many*. As Michael Gazzaniga explains:

The human brain contains myriad maps that correspond in an orderly way to the external world. Within the part of the brain governing our visual system there are at least thirty maps. Many of these areas of the brain are designed to handle such special visual tasks as motion, color, and shape detection. This parceling is evident from patients who suffer lesions in these various areas and afterward display strange abnormalities, such as their ability to detect color, shape and motion. With normal vision the neural processing that goes on in these separate areas is somehow combined so we can enjoy the conscious perception of a colorful moving object *(28)*.

A further level of organization, in the way the brain handles information, distinguishes detail from the big picture.

The perceptual world is organized hierarchically: the forest consists of trees, which in turn have leaves. Visual attention can emphasize the overall picture (global form) or the focal details of a scene (local components). Neuropsychological studies have indicated that the left hemisphere is biased towards local and the right towards global processing *(29)*.

The brain is constantly manipulating information and running parallel programs to maintain a milieu that allows it to continue its activities without interruption. A casual "good morning" appears so mundane that it might be regarded as entirely irrelevant to a consideration of how we use the faculty of reason. Yet the cerebral machinery employed for routine social interactions is, in essence, similar to what we use for playing chess championships or studying abstract mathematical theory. Brain power allows us to integrate an enormously complicated barrage of incoming information and formulate the most appropriate response.

13

MIND

The human mind is a device for survival and reproduction,
and reason is just one of its various techniques.
—EDWARD O. WILSON*

W HILE REASONING TAKES place in the brain, it also takes place in the mind. Where do minds exist in space and time? How does the mind interact with the brain and vice versa? Do animals have minds? The mind is our inside world, where we experience thoughts and emotions, where we perceive, believe, and recall, and where we build an image of the outside world.

We might expect to have a better understanding of the mind than we have of anything else, just because it is highly personal. Yet nothing could be further from the truth, for the mind is one of our deepest mysteries. The task of setting the mind to explain its own nature is like asking an eye to make a visual inspection of itself—there are difficulties. To explore the mind, the first question we shall ask is the oldest. How is the mind related to the body?

---

* This quotation is from *On Human Nature* (1978), for which Wilson received a Pulitzer Prize. He was born in Alabama in 1929. He pioneered the modern study of social behavior in animals, delineating the territory of a new discipline in his book *Sociobiology: The New Synthesis* (Cambridge, Mass.: Belknap Press of Harvard University Press, 1975).

The Mind-Body Problem

The mind-body problem has confounded thinkers for thousands of years, and the debate continues. Many disciplines have been summoned to find an answer—theology, philosophy, and more recently, psychology, physiology, and neurology. Perhaps the most eloquent of all discourses on the subject is the dialogue between Socrates and his friends the night before he was forced to drink hemlock as a death sentence. Plato recorded the discussion in the *Phaedo:*

> Let a man be of good cheer about his soul, who having cast away the pleasures and ornaments of the body as alien to him and working harm rather than good, has sought after the pleasures of knowledge; and has arrayed the soul not in some foreign attire, but in her own proper jewels, temperance, and justice, and courage, and nobility, and truth *(1)*.

Socrates believed we have the best chance of a healthy immortal soul if we dedicate our lives to philosophy.

The terminology has altered with time. The *Phaedo* shows how difficult it is to decide if an argument about the mind has reached a conclusion. How do we decide whether to accept the argument of Socrates? A. J. Ayer writes:

> The criterion which we use to test the genuineness of apparent statements of fact is the criterion of verifiability. We say that a sentence is factually significant to any given person, if and only if, he knows how to verify the proposition which it purports to express—that is, if he knows what observations would lead him, under certain conditions to accept the proposition as being true, or reject it as being false *(2)*.

Such high standards are not attainable when one is addressing the mind-body problem. Rather than give up on the issue, we may adopt the position of another philosopher, J. L. Austin, who suggested that a "bit of a criterion" of correctness would be "if you could get a collection of more or less cantankerous colleagues all to accept something after argument" *(3)*. But even this task is daunting when it comes to the mind-body debate.

THE IDEA OF an immortal soul, a form of persisting personal existence after death, started long before the invention of writing because early burial sites contained tools and ornaments intended for a life after death. Then the Greek philosophers codified the features of the soul. Socrates was emphatic that the soul was distinct from the body, pressing the point home in the *Phaedo:*

> "Consider, then, Cebes, if these are our conclusions from all that's been said: soul is most similar to what is divine, immortal, intelligible, uniform, indissoluble, unvarying, and constant in relation to itself; whereas body, in its turn, is most similar to what is human, mortal, multiform, non-intelligible, dissoluble, and never constant in relation to itself. Have we anything to say against these statements, my dear Cebes, to show that they are false?" "We haven't" *(1)*.

The essence of Socrates' argument was: (1) the soul has features that differ from those of the body, so the soul is a separate entity; (2) because the soul is a separate entity, its existence is not ended by death of the body.

Interestingly, these questions never arose in ancient Chinese philosophy, which was just as enlightened but perhaps more practical than its European counterpart. The notion of a soul that survived death was quite unnecessary for Confucius. He believed that the main concern we should have about death is the need to be remembered in a good light—and not to leave behind a metaphorical bad smell. Other cultures and religions of the time endowed ancestors with an afterlife, and proclaimed faith in notions such as reincarnation, but they did not emphasize any formal distinction between the mortal body and the immortal soul.

The Greek idea of juxtaposed body and soul became triumphant when it was adopted by both Christianity and Islam. Philosophers coined the term "dualism" for this body-soul relationship, and the concept has been entrenched ever since. Descartes was the most eloquent advocate for dualism.* He expressed his views in a series of elegantly framed arguments, and the issues he raised have been debated ever since. Descartes contended that we can be absolutely sure of the existence of our souls or "inner" worlds, because the act of thinking can take place only through the use of the machinery required for thinking—the soul. The soul's act of thinking proves its own existence; the proposition proves "I exist" is a fundamental truth because "I cannot make the assertion unless I exist." The primacy of this certainty overshadows his confidence in the reality of any "outer" world. "I can conceive of anything, apart from my own soul, as illusory."

Descartes concluded that the soul's irrefutable existence means that there is an essential difference between the "inner" world of the soul and the "outer" world of the body. This difference constituted grounds for making a fundamental conceptual separation be-

---

* René Descartes (1596–1650) is often regarded as the father of modern philosophy. He was educated in a Jesuit College at La Flèche and later studied law in Poitiers. He remained a Catholic all his life, though he must have experienced difficulty in reconciling some of the positions taken by the church, such as the Inquisition's handling of the views put forward by Copernicus, Bruno, and Galileo. He is reputed to have been serving in the army in 1619 when he had a vision, in a stove-heated room during winter, that the whole of knowledge could be restructured in a unified system of rational thought. His major works were *Discours de la méthode* (1637), *Meditationes de prima philosophia* (1641), and *Principia philosophiae* (1644). See E. S. Haldane and G. R. T. Ross, *The Philosophical Works of Descartes* (New York: Cambridge University Press, 1970).

tween body and soul. He took advantage of the knowledge of his time concerning the functions of the brain, and he focused attention on it having a special relationship to the soul. He considered that while the brain and the soul are distinct from each other, they can interact.

Descartes conceived of the soul as quite different from tangible objects. He took the opinions of Socrates and extended them. His analysis provided the ramparts of the modern dualists' stronghold. Dualists contend that human activities are directed by the soul, and our behavior is too complex to be accounted for in terms of physical mechanisms. Separation between body and soul can be inferred from the reports of people who have claimed to have seen themselves from outside their own bodies. The most attractive aspect of dualism is that it gives an explanation of how one might survive death. If the soul is distinct from the body, who cares if the body dies?

THESE ARGUMENTS FOR dualism present problems. In addressing them, we will employ the more modern term "mind" rather than "soul." The only difference between "mind" and "soul" is the attribute of immortality that several religions assign to the soul. Immortality, however, is not accessible to critical examination—we can believe in it, but we cannot confirm or refute it.

Taking the dualist points in turn, the first argument is challenging. "My ability to think proves the existence of my mind and I cannot be certain of anything else." But on critical scrutiny, all this really means is "I cannot imagine that my mind does not exist." In other words, "I can pretend that the outside world, including my body, does not exist; I can speculate on the possibility that everything but my mind is an illusion." But this does not mean that everything other than one's mind actually *is* an illusion. Is it reasonable to deny the widely held view that we live in a world that extends beyond our minds? Should we reject the existence of our bodies, explaining them in terms of mistaken perceptions? This po-

sition is difficult to sustain. The reality of the outside world becomes compelling when we are in danger or pain, when we are hungry or thirsty, and when we are sexually roused. From the viewpoint
of evolution, the outside world has been responsible for the way our
brains and minds developed.

There are also problems with the dualists' second assertion, that
the physical world is too simple to explain the intricacy of human behavior. Inanimate objects—computers—can easily surpass human
feats of arithmetic. Recently a computer has even beaten the world
chess champion.

The third argument concerns perceptions of being outside one's
body, but reports of disembodied experiences are, to say the least,
odd. Such experiences can be produced by psychotropic drugs and
by various psychiatric disorders. Highly emotional rituals can also
lead people into a communal feeling of being "outside one's body."
But these perceptions conflict with the normal, commonsense view
of reality, so in psychiatric terminology disembodied experiences are
hallucinations, and hallucinations cannot prove anything.

The fourth argument focuses on survival of the mind after death.
As previously discussed, immortality is not testable. Its existence
rests on religious authority. Reason cannot prove or disprove dogma.

There are more direct arguments against separating the mind
from the body. The essence of dualism is the assertion that the mind
has an existence independent of space and time, whereas the body is
locked into space and time; furthermore, dualism envisions an interaction between these radically dissimilar entities, mind and body. In
the words of Charles Sherrington, "Then the impasse meets us. The
blank of the 'how' of mind's leverage over matter."* The converse,

---

* Sherrington's work on neurophysiology culminated in his classic monograph *The In-
tegrative Action of the Nervous System* (1906). The quotation comes from a more philosophical book, *Man on His Nature* (New York: Macmillan and Cambridge University
Press, 1941), based upon his Gifford Lectures delivered at the University of Edinburgh in
1937 and 1938. Sherrington received the Nobel Prize in 1932.

"matter's leverage over mind," is, of course, equally problematic. General experience within our universe suggests that physical events cause physical events, yet dualism claims that the nonphysical mind commands the physical body. No explanation is offered for how this might occur. Taking all the arguments into consideration, the case for dualism looks decidedly insecure.

IF DUALISM IS rejected, what are the alternatives? One possible resolution of the problem is that the mind is simply an expression of the functioning of the brain. Can we be looking at the same thing—mind and functioning brain—from different viewpoints? Neurophysiological studies show the existence of a time lag between electrical events taking place in the brain and mental events taking place in consciousness, but this does not mean that they are inherently separate. A clap of thunder occurs before the pressure waves reach us, yet the original thunderclap is the same pressure wave.

The mind cannot be the only result of activity in the brain, for we know the brain is busy controlling bodily functions of which we are completely unaware. Furthermore, electrical activity of nerve cells continues during states of coma, where, by definition, mental activity is replaced by a state of deep unconsciousness. The question must therefore be framed more precisely. Can mental function simply be one category of brain activity? In this way, every event in the mind becomes identical with a physical event in the brain. Whether such phenomena are described in terms of the mind or the brain would depend upon the context. A garment can be looked upon as an object to keep us warm, convey social status, attract the opposite sex, confer group identity, and so on. In other words, it has several different functions, but it is still just a garment.

SO A POSSIBLE interpretation of the evidence is that brain activity and mental phenomena are simply different descriptions of the same thing. As Spinoza put it, "Mind and body are one and the same individual" *(4)*. Alexander Bain asserted the same opinion with stern simplicity: "Many persons, mocking, ask—What has mind to do with brain-substance, white or grey? Can any facts or laws regarding the spirit of man be gained through a scrutiny of nerve-fibers and nerve-cells?"* Later he answers: "The arguments for the two substances [brain and mind] have now entirely lost their validity. The one substance with two sets of properties, a double-faced unity, complies with the exigencies of the case." This "functionalist" position is, perhaps, the least unsatisfactory resolution of the mind-brain problem.

CONSCIOUSNESS

What part of the brain is concerned with the consciousness? Much has been written on this topic, but we have no answer. The more recently evolved areas of the superficial layers of the cerebral hemispheres play some role, but not an exclusive one, in conscious processes. Different parts of the brain are concerned with different kinds of mental function, but we cannot detect any special characteristic distinguishing the *nerve impulses* concerned with consciousness; nor is there evidence that the *nerve cells* involved in conscious processes possess any unusual features.

When we are awake, electrical activity from the brain, recorded over the head, is complex, and of low amplitude. When we are asleep, the pattern turns into slow oscillations of high amplitude. As drowsiness merges into sleep, the electrical activity changes gradually, with no electrical indication of the point at which consciousness is lost. The higher-amplitude, slower oscillations of electrical activity increase as sleep becomes deeper. When we dream, the pattern reverts from that

---

* With these words, Bain opened his treatise *Body and Mind: The Theories of Their Relation* (1873). The son of a weaver, he became professor of logic and rhetoric at Aberdeen University.

of deep sleep toward the lower-amplitude, faster activity seen in drowsiness and light sleep; in addition there are rapid eye movements.

Sleep imposes itself, quietly but firmly, with minimal disturbance beyond the irritation of wasting our time. "We sleep, but the loom of life never stops and the pattern which was weaving when the sun went down is weaving when it comes up tomorrow."* Sleep must serve some essential biological function, for most of us spend some 25 percent of our life asleep—during which our ancestors would have been particularly vulnerable to predators. Time spent asleep is also time lost from other, seemingly more productive activities such as seeking food. We now have at least a partial answer to the puzzle of why we sleep, for nerve cells manufacture proteins much faster during sleep, and the rate increases in deeper sleep *(5)*. The accelerated production of proteins takes place in spite of a reduction in energy consumption of some 25–30 percent—sleep "charges the batteries" of the brain *(6)*.

When we wake from sleep, we regain consciousness. We take this for granted, but consciousness is poorly understood. William James described the stream of consciousness in a graphic analogy: "Like a bird's life, it seems to be an alternation of flights and perchings" *(7)*. More recently, Edward O. Wilson went further by linking mental and physical occurrences in the brain:

> There is no single stream of consciousness in which all information is brought together by an executive ego. There are instead multiple streams of activity, some of which contribute momentarily to conscious thought and then phase out. Consciousness is the massive coupled aggregates of such participating circuits *(8)*.

---

* From Henry Ward Beecher (1813–1887), an American clergyman and writer who preached for temperance and denounced slavery. His church in New York raised and equipped a volunteer regiment for the Civil War. His aunt was Harriet Beecher Stowe.

In analyzing the neurophysiology of a simple "good morning" greeting, we saw how events in the outside world generate specific patterns of nerve impulses in specific areas of the brain. The biological importance of the event, together with its context in past experience, determine whether the pattern of nerve impulses in the brain enters consciousness as a perception.

Paul Churchland has addressed the puzzle of conscious experience without external stimulation. "Upon first reflection, self-consciousness is likely to seem implacably mysterious and utterly unique." He answers himself: "self-consciousness involves the same kind of continuously updated knowledge that one enjoys in one's continuous perception of the outside world" *(9)*. Consciousness without perception derives from processes in the brain that are similar to those generated by an external stimulus—but the relevant patterns of nerve impulses originate within the brain, perhaps by the repeated monitoring and retrieval of information stored in memory.

RECENTLY, VARIOUS METAPHORS have come into vogue to help explain consciousness. Francis Crick has suggested that it is generated by a "searchlight of attention" between the thalamus (a group of nerve cells at the base of the brain) and the cerebral cortex.

> What do we require of a searchlight? It should be able to sample activity in the cortex and/or the thalamus and decide "where the action is." It should then be able to intensify thalamic input to that region of the cortex, probably by making the active thalamic neurons in that region fire more rapidly than usual. It must then be able to turn off its beam, move to the next place demanding attention, and repeat the process *(10)*.

This concept is also used for a metaphor in which the searchlight is part of a larger model, namely a theater *(11)*. Information about the outside world is brought in from diverse sources to be integrated in a theatrical production. The "searchlight" becomes a spotlight that focuses attention on the center of action on the stage. The illuminated image is seen by the audience, which is made up of all the unconscious elements of mental function. This notion is controversial, for while some cognitive scientists consider that there is a single area in the brain where consciousness comes together, others remain skeptical. However, most agree that consciousness can handle only one task at a time, whereas unconscious mental activity can run many tasks simultaneously. In computer jargon, consciousness displays serial processing, whereas the unconscious mind is capable of parallel processing.

FROM THESE THEORETICAL deliberations, we must conclude that there is no coherent explanation of the physical basis of consciousness, although lively erudite debates now range over philosophy, psychology, and neurophysiology.* There have been developments in many areas of cognitive science where objective measurements can be made. We can record nerve impulses, trace the pathway of nerve fibers, and measure the release of neurotransmitters and the distribution of their receptors. We can also measure the physical and mental reactions to various stimuli. But the pivotal question concerns how a physical event in the brain leads to a conscious event in the mind, and here we have nothing approaching an answer.

We have argued that provided there is a suitably receptive background of brain activity, stimulation of a sense organ elicits a specific pattern of nerve impulses that *is* the perception. We cannot explain how a pattern of nerve impulses becomes a conscious phe-

---

* For recent discussions, see David J. Chalmers "The Problems of Consciousness" and Patricia S. Churchland, "What Should We Expect from a Theory of Consciousness?" These papers pose many questions and acknowledge that we have few answers. They are published consecutively in "Consciousness: at the Frontiers of Neuroscience," *Advances in Neurology,* vol. 77, pp. 7–32, eds. H. Jasper, L. Descarries, V. F. Castellucci and S. Rossignol (Philadelphia: Lippincott-Raven, 1998).

nomenon, and we cannot predict when this problem will be solved. We cannot even be sure that we will ever be able to understand the mechanism in the same way that we understand how the heart pumps blood, the stomach digests food, or the kidney excretes urine. Insights might arise from some radical advance in our knowledge of physics, chemistry, and neurophysiology, but, equally, we may simply be unable to crack the problem. We may be up against the limits of our mental machinery, just as physicists seem to be when confronted with the question of what existed prior to the "big bang" 10 billion years ago.

THE MINDS OF ANIMALS

Over a century ago, George Romanes argued that since animals frequently behave like humans, we can infer they have mental states in some way comparable to our own. Romanes acknowledges that it is impossible to prove the point, but

> common sense, however, universally feels that analogy here is a safer guide to truth than the sceptical demand for impossible evidence. . . . common sense will always and without question conclude that the activities of organisms other than our own, when analogous to those of our own which we know to be accompanied by certain mental states, are in them accompanied by analogous mental states* *(12)*.

This argument poses some difficult questions. Where do minds start to appear in evolution? The answer must depend on how "minds" are

---

* Romanes was a close friend of Charles Darwin. He was born in Canada and came to England at a young age, where he published a book entitled *Animal Intelligence* in 1881. He supported Darwin's ideas in *Mental Evolution in Animals* (1883) and *Mental Evolution in Man* (1888). Romanes became zoological secretary of the Linnean Society. His accounts of astounding mental feats by cats and dogs do not stand up to critical scrutiny, but his general conclusion is challenging.

defined. There is another more profound problem—we cannot get into the mind of an animal because we cannot begin to imagine what the inner world of an animal is like. It is difficult enough to understand other people's minds, so how can we study the minds of animals? Certainly we must take care not to attribute human mental states too easily to animals, but the essence of Romanes's argument has a persuasive ring. If we accept that the diversity of species is achieved because each has developed effective ways of coping with the environment, we must acknowledge that minds are certainly useful instruments for surviving—they bring biological advantages just like hair and teeth. And just as hair and teeth evolved in similar but distinct ways in different species, so, we might surmise, have minds.

While verbal language has made it much easier for us to manipulate and communicate our mental processes, neurological observations show that mind and language can be separated, so it is no longer possible to argue that mind can exist only where there is language. Indeed, we can often tell more about what other people are thinking by how they look than by what they say. Nevertheless, our minds would be slower, more restricted, and harder to understand if we could not interpret, store, retrieve, and manipulate words. So we should not be surprised by our difficulty in exploring the minds of speechless animals.

ANOTHER SCHOOL OF thought, in comparative psychology, takes a much more objective stance on the analysis of animal behavior. Analogies to human mental states are deplored, and new words are coined to underscore the distinction between mental and physical processes. For example, in order to avoid any possible confusion with human experiences, the terms "hearing" and "sight" are replaced by "phono-reception" and "photo-reception." This approach is extreme, and although we cannot criticize its rigor, its results have been disappointing. Strict objectivity makes it difficult to ponder the most interesting question we can ask of comparative psychology: What

can we learn from animals that might help us to understand ourselves?

We shall therefore take a position that seems probable but not definite. Animals have minds, but they are not readily accessible to
us. We cannot study the minds of animals directly, yet the more
recently evolved mammals are likely to feel pain, pleasure, anger,
fear, and affection. John Locke drew the same conclusion: "in all the
visible corporeal world we see no chasms or gaps. . . . down from us
the descent is by easy steps and a continuous series of things, that in
each remove differ very little one from the other. . . . There are some
brutes that seem to have as much knowledge and reason as some that
are called men" *(13)*. Locke describes a steady continuum and no line
can be drawn to show exactly where mind appeared; it evolved gradually, with increases in the size and intricacy of the brain. Frans de
Waal has forcefully reasserted this argument in a contemporary setting *(14)*.

Animals can be trained to find their way through mazes and to
press levers to obtain food. These patterns of behavior can be interpreted in more than one way, but at some level, perhaps in monkeys
and apes, the attribution of intention seems appropriate. Intention,
however, cannot be proved—it can only be inferred. The difficulties
in comparing the minds of animals and humans will never disappear, but the evidence that other primates have minds is so persuasive that Ramón y Cajal felt compelled to ask how monkeys might see
us: "Described by the monkey, what would man be? Probably a sad
case of degeneration, characterized by a contagious mania for talking and thinking" *(15)*.

MODELS OF THE MIND

Another way to study mental function is to construct a model of how
the mind might work. Colin McGinn has suggested a model with
mental phenomena classified into two broad categories: "sensations"
and "attitudes" *(16)*. His concept of "sensations" includes perceiving
the outside world, feeling emotions, and experiencing bodily func-

tions. His concept of "attitudes" includes thinking of propositions, such as whether it is too cold to go outside, and considering objects, such as a hat or a coat. Mental phenomena *may or may not enter consciousness*—for example, we may or may not be aware of all our "attitudes"—but all mental phenomena are capable of entering consciousness. An event, a process, or a state is mental *if, and only if, it is or can be conscious.* In McGinn's model of the mind, "sensations" may result in immediate actions through the need to satisfy drives, but for much of human behavior, "sensations" lead to the formation of "attitudes" which, in turn, determine what we do.

While McGinn's concept of "sensation" includes "perception," some psychologists and philosophers have suggested that the term "sensation" should be confined to direct, crude experience, whereas "perception" should be used for the more refined mental constructs that we assemble and interpret as images of the outside world. For example, a dog generates a range of individual "sensations" that impinge on our inside worlds: his color, his linear and shaded visual contours, his bark, his smell, the coldness of his nose, and the texture of his fur. In addition, we regard dogs as animals in our outside worlds—we have "perceptions" of their existing externally and quite independently of us. "Sensations" have a raw, immediate nature; "perceptions" tend to be shaped by our previous experience. We may "perceive" a dog as a friend, a hunting companion, a guard, or a potential meal.

Do we create "perceptions" from "sensations," or are they parallel categories of experience? As yet we have no answer. Some argue that we assemble "sensations," in our conceptual framework of space, time, and causation, to produce "perceptions; others disagree, citing evidence from patients with selective disruption of "sensation" or "perception" *(17)*.

BELIEF AND DESIRE occupy positions of central importance as "attitudes" in McGinn's model of the mind. The terms "commonsense psychology" and "folk psychology" have been coined for a simple model crafted from belief and desire, to give a rough-and-ready explanation of how our minds work. According to this model, all mental activity breaks down into steps that are either beliefs or desires. The core of commonsense psychology is the generalization that we desire certain ends, and we believe that we can achieve them by taking suitable actions. Sometimes we are aware of the beliefs and desires directing our actions, but not necessarily. To illustrate the operation of commonsense psychology we can look at a simple fishing trip. I have heard that there are large trout in a local river (the belief). With this background, I develop a wish to go fishing (the desire), so I search for my rod (the suitable action). I wish to have company (the desire), and I think my son would enjoy it (the belief), so I invite him (the suitable action).

The belief-desire principle offers a general explanation of behavior, and by attributing belief and desire to other people, we can often predict what they will do. How can we apply commonsense psychology to the task of tracing the origins of reason? When agriculture began, farmers with sharp minds saw that vegetation grew better near water, so water led to thriving plants (the belief). They wanted to increase their crops (the desire), therefore they sowed their seeds along the riverbanks (the suitable action). We can see how the simple components of belief and desire can be assembled into something that looks like an embryonic faculty of reason.

Other models have been proposed to explain how the mind operates. Freud, Jung, and Adler held sway for many years. Freud is the most widely read and quoted; from his clinical experience as a psychiatrist, he concluded that cultures establish taboos which modify the expression of certain powerful instinctive drives, the most notable being sexual. He argued that the conflict between an individual's instinctive drives and social constraints leads to psychiatric symptoms. Furthermore, these symptoms appear without any awareness of their cause.

Freud displayed moral courage in presenting his views when society was straitlaced and hypocritical. Unfortunately, his insights led him to propose a model of the mind that grew increasingly complicated and top heavy. A theoretical construct was erected and buttressed by such hypothetical notions as the id, the ego, and the superego. During the last thirty years the edifice cracked and now the entire structure is falling apart.

KNOWLEDGE AND MEMORY

Exploration of knowledge and memory leads us back to the functioning of the brain, in particular the correlation between neurological symptoms and brain damage. Diseases may cause widespread damage in the brain, as with multiple sclerosis, or the damage may be confined to a circumscribed region, as with a stroke. For more than a century, pathologists have been collecting evidence that has allowed them to correlate disturbances of neurological function, recorded during life, with the focus of brain disease, recorded after death.

Conclusions drawn from these correlations have been confirmed by observations made during neurosurgical procedures. When operations are undertaken for intractable epilepsy, the surgeon often applies weak electrical shocks to the surface of the brain to obtain a navigational fix on the map of anatomically localized cerebral functions. Stimulation causes an effect determined by the circuitry of the underlying nerve cells and fibers. Mapping the surface of the cerebral hemispheres in this way has shown that distinct areas are responsible for vision, hearing, speech, skin sensation, and the execution of movement. Furthermore, within a field dedicated to a particular function, such as voluntary movement or skin sensation, there is a mosaic of overlapping regions, each of which is responsible for a different part of the body.

What is the significance of this anatomical and physiological organization? Fifty years ago an analogy was drawn between nerve fibers and telephone lines. Now this approach is regarded as oversimplified, because the brain employs parallel processing and modular circuitry that cannot be matched by any telephone system. Nevertheless, there are some similarities between the brain and a telephone system. Telephone lines are widely distributed, but they come together at certain locations where redirection services are concentrated—what used to be called telephone exchanges. The telephone exchanges have a special role as centers for integration of information. Destruction of a telephone exchange will have more effect on the operation of the service than damage elsewhere. This analogy illustrates how cerebral localization is the result of high concentrations of impulse traffic serving particular functions.

As the working of the human brain was being unraveled, similar investigations on animals were running into difficulties. Instead of simplifying the problems and helping the analysis, animal experiments seemed to put up roadblocks. K. S. Lashley studied the effect of cutting connections in the brains of rats after they had been trained to negotiate a maze.* He then measured the number of trials required to relearn the maze. He found that the extent of retraining depended on the amount of damage rather than its site. Then evidence more in keeping with the human studies began to appear—memory is initially stored in the specific areas of the brain that are also involved in processing incoming impulses. The paradox in Lashley's observations could now be explained, because the ability to run a maze requires many categories of information and so many neural pathways are involved. The consequent diffuse engagement of the brain, for establishing memory, and running a maze explains why Lashley failed to find localized storage of memory.

A philosopher made the next important step toward understand-

---

* K. S. Lashley was born in Virginia in 1890. His classic monograph was *Brain Mechanisms and Intelligence* (Chicago: University of Chicago Press, 1930).

ing the nature of knowledge and memory. Gilbert Ryle drew a shrewd distinction between "knowing how" and "knowing that."* "Knowing how" to play the violin is a learned skill, whereas "knowing that" the violin was made by Stradivarius is a learned fact. There are other distinctions between "knowing how" and "knowing that." We can believe a fact, but we cannot believe a skill. We can perform a skill, but we cannot perform a fact. In addition to these differences, there are similarities between "knowing how" and "knowing that." For example, it is meaningful to remember both a skill and a fact—in each case what has been learned can, in some sense, be regarded as knowledge.

REASON IS A special method of processing previous experience that has been stored in a "data bank" of the mind called memory. To analyze the nature of reason, we must therefore seek an understanding of memory. Loss of factual memory has been associated with lesions involving (1) the lower and inner aspect of the temporal lobe (the hippocampus) and overlying temporal cortex—areas called, for short, "the temporal complex"; and (2) the frontal cortex and its connections with the mammillary bodies and the dorsomedial thalamic nucleus—areas called, for short, "the frontothalamic complex." Patients with damage in the temporal complex or the frontothalamic complex lose the ability to learn and recall facts, but they can still learn and recall skills *(18)*. The regions responsible for remembering skills are older in evolution—the basal ganglia and the cerebellum. These structures are also responsible for the planning and coordination of movement.

Short-term factual memory is stored within an ensemble of those regions of the brain responsible for the initial, primary processing of

---

* Gilbert Ryle was professor of philosophy at Oxford and editor of the journal *Mind*. His book *The Concept of Mind* (1949) introduced many of the ideas that are now regarded as fundamental for modern psychology. It was first published in 1949 (London: Hutchinson).

the various types of sensory information—sight, sound, smell, touch, or pain. For the consolidation and retrieval of long-term factual memory, the temporal and frontothalamic complexes interact with the widespread storage sites in the primary processing areas of the brain. (18)* In addition to this quite elaborate anatomy of memory, areas of the cerebral cortex at the front of the brain integrate past experiences with present circumstances in a continually active "scratch pad" for short-term use.

Recent studies have added yet another layer of complexity to the mechanism of remembering—the existence of *explicit* memory accompanied by conscious awareness, and *implicit* memory that is acquired without intention and sustained without knowledge *(19)*. Seeck and colleagues have shown pictures of faces to people while recording electrical brain activity *(20)*. They found a different distribution of responses according to whether the pictures had been shown previously. These differences occurred even if the subject did not remember whether the face had been shown before. Thus visual information must undergo parallel processing in the brain, and from the distribution of the responses it seems that explicit memory is handled in the lower and inner aspect of the temporal lobe, whereas both implicit and explicit memory involve association areas, such as the frontal lobes. Perhaps implicit memory is available for quick, automatic reactions to the environment—we can even extend this speculation and suggest that implicit memory is the "data bank" for the unconscious mental processes that can, under suitable circumstances, force their way up into consciousness.

OTHER ASPECTS OF memory remain even more obscure. The capacity for memory must be finite, so organized forgetting serves

---

* This activity must be sustained for many years, and eventually the assembled information becomes independent of the temporal complex but not the frontothalamic complex.

a necessary biological purpose, though we know very little about it. We do not understand how memory is encoded, stored, or retrieved at the cellular or subcellular level. We do not even know whether the nerve cells use chemical or electrical mechanisms. Information could be coded in the structure of proteins; in this case acquisition and deletion of memory would be achieved by rapid modification of molecular structure. Alternatively, information could be stored in the dynamic interaction of impulse traffic among networks of nerve cells, just as the active circuits of a computer can keep data before saving them on magnetic media. The complexity of the entire memory system is striking but not exceptional. A similarly elaborate orchestration of brain activity underlies voluntary movement, language, and in all probability, reason.

MOTIVATION

We have already given separate consideration to the elements of motivation—instincts, emotions, and cultural imperatives all have the ability to drive us because they invoke mental rewards and punishments. How do the brain and the mind create rewards and punishments? Neurophysiological studies show that "feeling good" is associated with activity in a processing center that releases the neurotransmitter dopamine in small groups of nerve cells beneath the cerebral cortex—these make up the ventral striatum, comprising the nucleus accumbens and the extended amygdala (21). Various pathways involved with rewards use a variety of neurotransmitters, but they seem to converge on the dopamine system projecting to the ventral striatum (22). From here, impulse traffic diverges and ultimately reaches the cerebral cortex, particularly the frontal and temporal areas.

We can assemble a speculative general hypothesis that satisfaction of drive leads to increased impulse traffic feeding via the ventral striatum into the cortex, and if the excitation is sufficiently intense there is microsynchronization. This scheme is in keeping with the evidence that mystical states, aesthetic experiences, emotions, and orgasms are all associated with microsynchronization in various parts of the brain. It offers a working hypothesis, but nothing more until further evidence becomes available.

FROM THIS THEORY of motivation, we can go on to consider how drives are directed and redirected. As we have discussed, instinctive drives are generated from an internal state (substances circulating in the blood and genetic predisposition) interacting with an external stimulus (triggering the onset of the response). Emotions provide motivation because they entail needs that crave satisfaction. Hume saw how passion, a blend of instinct with emotion, could dominate reason. He wrote: "Reason is, and ought only to be the slave of the passions, and can never pretend to any other office than to serve and obey them. . . . Reason alone can never be a motive to any action of the will" and "it can never oppose passion in the action of the will *(23, 24)*." Nevertheless, Hume would certainly have upheld the usefulness of reason, for many things can be done with a slave that cannot be done without one.

Raw instinct and emotion must have dominated the lives of our earliest hominid ancestors, but, for later hominids, cultures have controlled behavior by modifying the expression of basic biological drives. How is this accomplished? We learn emotive cultural attitudes, that set our goals, just as we learn everything else. First, operant conditioning comes into play, with mental rewards (positive reinforcement) in the shape of social respect and mental punishments (negative reinforcement) in the shape of social disdain. Second, there is modeling—children copy their brothers and sisters, their parents and their teachers. In addition to these mechanisms, that are shared by most mammalian species, we have a unique, highly versatile, and exceptionally powerful technique—the telling of stories. Stories provide an enormous expansion to the repertoire for modeling, extending it as far as the imagination can go. People identify with the heroes and heroines and, most importantly, "the appropriation of culturally circulated stories results in the adoption of the stories' goals" *(25)*.

In addition to fashioning moral attitudes, narratives can convey useful facts concerning the nature of the world. Xenophon gave information on horsemanship, hunting, horticulture, and how to manage estates. It is very likely that our earlier ancestors passed on simpler narrative tips, from one generation to another, on how to get food, how to make shelters, and how to light fires.

While narratives can reinforce morality and become repositories of the culture's accumulated knowledge, they can also serve another function—they entertain us. Stories have a power to capture our attention quite apart from their moral and factual content. They conjure up pictures of the world in time and space, where events have causes but are not constrained by reality. How is this imaginative exercise capable of tapping into our emotions and influencing our attitudes? What matters is how the "message" of the story—the perspectives and the aspirations—fits in with our view of the world. This view, in turn, is the result of the stories we have heard or read from the time of our youngest childhood. The whole fiction industry is built upon our curious predisposition to become enthralled by stories that we know to be untrue, if they reflect our cultural cosmology. To accept stories outside this framework, we have to undergo "brainwashing."

Why are we gripped by what we know to be untrue? It is disconcerting to become so engaged in something that seems to have no purpose. We cannot have acquired such susceptibility unless it has biological value. How did our capacity for telling and listening to stories arise over the course of evolution? We do not have to look far to find an explanation. The oldest stories are religious, so narrative, like art, is derived from religion. Indeed, stories are not simply like art, they *are* art. Depending upon their content, and the skill with which they are crafted, stories can bring us aesthetic experiences. Stories can carry us into another world where we may cherish serenity, find excitement, laugh, or just escape from our daily routine. Narrative messages can also be combined with music to gain special effects, through chants, hymns, national anthems, operas, and even advertising jingles and popular song.

Placing all these facets in perspective, stories emerge as a powerful means to frame our cultural attitudes. If we share in a collective imagination, we share an outlook—the biological purpose of fiction is cultural—to mold our feelings and unite our views. Stories build cultural consensus and reinforce cultural solidarity.

Reasoning

Reason is built upon a platform of logical induction (observations lead to conclusions which allow predictions) and logical deduction (if a and b are two classes and a is contained in b, then x is in a implies that x is in b). Reason assigns priority to observation over theory (Galileo's knife) and simplicity over complexity (Ockham's razor); it also demands consistency, coherence, and efficiency.

With this background, the study of reasoning has focused on the process by which we make decisions, and here an unexpected but consistent theme has been the uncovering of emotional elements enmeshed with reason, without our being aware of them. Antonio Damasio has recently reported some intriguing evidence of emotion entering unnoticed into the process of making decisions. An electrical test, termed the skin conductance response, is generally accepted as signaling a variety of emotional reactions, and the test is sensitive enough to detect subtle responses that are insufficient to enter consciousness. In normal subjects who are making decisions under experimental situations, skin conductance changes occur before conscious choices are made. Patients with frontal lobe lesions do not have these preliminary skin responses, and they have difficulty in making appropriate choices. Damasio argues that the skin responses

signal a brief, early stage of unconscious bias which sets the background and influences the result of our conscious process of decision
making *(26)*.

D. F. Halpern illustrates the role of cultural influence on whether
we accept or reject an argument: "Some women who have abortions
are remorseful. Some remorseful women are psychologically disturbed. Therefore some women who have abortions are psychologically disturbed" *(27)*. While each premise can be confirmed by
observation, the conclusion is false because the argument is flawed;
it is entirely possible that none of the women who had abortions and
were remorseful fell into the category of being psychologically disturbed. Nevertheless, the conclusion is likely to be accepted by people who are opposed to abortion.

Tversky and Kahneman have gone further in studying the influence of irrational attitudes in their paper "The Framing of Decisions
and the Psychology of Choice" *(28)*. They gave questions to students
at Stanford University and the University of British Columbia, and
they found that they could reverse the answers by recasting the
questions in a different form. They took a hypothetical task of
choosing between two options. An outbreak of viral disease is predicted to kill 600 individuals if there is no intervention. Option A
would definitely save 200 people. Option B would have a one-third
probability of saving all and a two-thirds chance of saving none.
Most students chose option A, although both options would save
the same average number of people (200) over a series of trials. It
just seemed better to avoid any risk and be sure of saving 200. The
question was then reframed and presented to another group of students who were asked to choose between option C, which would result in the certain death of 400 people, and option D, which would
have a one-third probability of killing nobody and a two-thirds
chance that everyone would die. Here, the majority chose option D,
because the possibility of saving 200 seemed better than the certainty of killing 400.

So our decisions are influenced by unrecognized and uncontrolled factors. We favor certainty if we are dealing with a potential gain and uncertainty if we are facing a potential loss. These attitudes are emotional—we minimize our anxiety by making decisions in this way. The emotional need to reduce anxiety is a secondary goal that provides a shortcut when the primary goal is too difficult. In a sense, we cut corners by making emotionally comfortable choices, and our cultural upbringing will contribute to what is emotionally comfortable. From an evolutionary standpoint, this mechanism must have served a useful purpose in situations where there was no time to ponder, and even now, some of the attitudes that we have absorbed from our culture stand us in good stead.

Emotional responses set by our culture can also be deleterious. They operate beneath the surface of consciousness, and they will influence decisions that might have been reached by reason alone. Scientists are as vulnerable to these irrational influences as anyone else. While the methods of science can often generate a clear answer to questions, the evidence is sometimes inconclusive. Scientists are not deterred; they still draw conclusions in these gray areas. At a recent meeting of the American Association of Physical Anthropologists, those present were polled on the vexed question of whether early hominids were initially hunters or gatherers. After voting, the same members were asked if they voted Republican or Democratic in the last presidential election. The answers were analyzed, and a high correlation was found between Republican voters and the hunter hypothesis; to the same extent, Democratic voters favored the gatherer hypothesis. Attitudes can also sway the making of decisions when the evidence *is* conclusively clear. Stuart Sutherland gives a long list of irrational decisions by wise members of society—legislators, judges, lawyers, generals, economists, scientists, and physicians: "the real proof of the prevalence of irrationality comes from the massive amount of research on the topic undertaken over the last thirty years by psychologists" *(29)*. This research is not widely known because there is no public interest in it. We take our inner worlds for granted, and we assume that we know all we need to know about them—a curious contrast to our fascination with the outer world.

Another example of the influence of attitudes on decisions was poignantly demonstrated in an experiment conducted by Stanley Milgram (30). He investigated the power of authority—an important issue during the last sixty years, over which "obeying orders" has been a consistent defense in trials for crimes against humanity. Milgram studied the behavior of an individual whom we shall designate as the "operator." The director of the investigation told the operator to give increasingly severe and painful electric shocks to the subject of an experiment on memory. The experiment was fictitious and the shock machine did not really generate electricity, but the operator did not know this, and the subject receiving the "shocks" behaved in every way as if he was receiving a painful stimulus. The director told the operator to increase the strength of the shocks in spite of the apparent distress of the subject. Variations were studied whereby the personality and status of the director, operator, and subject were changed. There were even observations on the effect of manipulating the relationship between the operator and the subject, for example exploring the impact of prior social contact between them. The results were disturbing: the operators did whatever they were told, whatever the consequences. The term "destructive obedience" was introduced to describe this disquieting phenomenon.

BEFORE WE LEAVE the topic of how we choose between options, mention must be made of the prisoner's dilemma, a predicament that serves as a model for how we approach many of the options in our complex social life when we do not know the intentions of others. The problem was first expressed as the choice facing two prisoners who are accused of committing a crime together and who cannot communicate with each other. They are given the following options:

1. If one confesses and the other does not, the former goes free and the latter gets 10 years of imprisonment.
2. If both confess, they both get 2.5 years.
3. If neither confesses, they both get 1 year.

The best joint outcome is reached if neither confesses, for this results in the lowest total punishment. But not confessing carries a high risk, for it leads to the largest individual punishment if the other prisoner confesses. Situations similar in principle to the prisoner's dilemma occur every day. We try to deal with these by seeking shared attitudes—some consensus on an approach. Thomas Schelling gives an example:

> When a man loses his wife in a department store without any prior understanding on where to meet if they get separated, the chances are good that they will find each other. It is likely that each will think of an obvious place to meet, so obvious that each will be sure that the other will be sure that it is "obvious" to both of them *(31)*.

As Schelling points out, the husband does not simply predict where his wife would choose to go; he has to predict where his wife would predict that he would choose to go, and so on ad infinitum.

So we see that reason is frequently engaged to help us deal with day-to-day choices where we have inadequate information, and the commonest of these is the task of interpreting other people's minds. Ironically, it seems that while reason evolved for this very purpose, all too often we are unable to predict what other people will do. In contrast, reason has been highly successful in helping us understand the nature of the physical world.

DISORDERS OF THE MIND

Disruption of the mind can be caused by neurological or psychiatric illness. The separation between psychiatry and neurology was a by-product of dualism; it grew from the expectation that psychiatry would deal with diseases of the mind, while neurology would deal with diseases of the brain. We now know that the mind and the brain cannot be separated in this way. Neuropharmacology has shown that the commonest serious psychiatric disorders, schizophrenia and depressive illness, are both associated with changes in brain chemistry. In schizophrenia, excessive transmission by dopamine takes place, so drugs that block dopamine are the cornerstone of treatment. The mechanism of depression is less clearly established, but there seems to be inadequate transmission by noradrenaline and serotonin; accordingly, drugs that increase the effects of noradrenaline and serotonin are the standard treatment. Neurological disorders such as Parkinson's disease and epilepsy are also treated by manipulating neurotransmitters. Thus the conceptual barriers between psychiatry and neurology have crumbled.

In schizophrenia, the mind becomes consumed by hallucinations, delusions, disturbances of thought, and ultimately disintegration of the personality. In depressive illness, the mood becomes dominated by melancholia, guilt, and often agitation, but the personality usually remains intact. The third common disease to attack the mind, Alzheimer's disease, has traditionally been regarded as a neurological illness because the brain shrinks and there are characteristic abnormalities in the nerve cells, readily seen through the microscope. In Alzheimer's disease the intellect degenerates and large numbers of nerve cells die. Many of the surviving nerve cells have abnormalities called tangles, and groups of damaged nerve cells aggregate to form plaques. Destruction involves the cortical regions that have abundant interconnections—the association areas. A less common dementing disorder is Pick's disease; this is of interest because it primarily attacks the frontal region of the brain.

In general, the dementing diseases lead to deterioration in memory, language, and reasoning, and the portions of the brain that are predominantly affected are those with a role in processing information. These pathways link up different groups of nerve cells within the brain, rather than receiving information from the outside or sending information to the outside. Dementing illnesses tend to spare vision, hearing, taste, touch, level of consciousness, and control of movement.

Humphrey has argued that disease confined to certain regions of the brain can lead to impairment of *perception* in patients who still have normal *sensation*—information is received, but it cannot be interpreted *(17)*. A patient with this kind of disturbance is unable to perform a task that requires an *explicit* function that has been lost. However, the patient can still pursue activities that make *implicit* use of the same function. For example, L. Weiskrantz has described a phenomenon termed "blindsight," in which patients lose their vision because of damage to the area of the cerebral cortex that receives input of visual information *(32)*. Such patients deny seeing a bright light in their lost visual field, yet they can consistently point to it. Without any conscious visual experience, such patients say that they are merely guessing. Similar dissociation between explicit and implicit loss of function occurs in patients with disorders of language (aphasia), reading difficulties (dyslexia), inability to recognize familiar concepts (agnosia), and loss of awareness of one side of the body (hemineglect syndromes).

Do any disorders attack reason selectively? Can they tell us what parts of the brain are responsible for reasoning? Special responsibility for reason seems to reside in the association areas of the cerebral cortex, which are richly endowed with interconnecting pathways linking different regions of the brain. The frontal areas underwent the most rapid enlargement as reason evolved in hominids, and disease confined to the frontal lobes causes the dysexecutive syndrome—difficulty in assembling the sequences of thought required for rational decisions.*

* See page 16.

14

EPILOGUE

The truth is rarely pure and never simple.
—OSCAR WILDE*

W E S T A R T E D O U R search for the nature of reason by defining it and tracing its origins through evolution. We then looked at how reason influences our lives, and we placed it in a historical perspective. We can now ask where all this has led us.

We have attempted to assemble pieces of evidence that can be fitted together like an incomplete jigsaw puzzle. Some conclusions can be drawn from the recognizable fragments of the picture, but these will have to be modified as new pieces of the puzzle are found. From what we can see at present, reason is a biological instrument— a highly versatile instrument, but still just an instrument, without any independent, direct ability to generate, replace or refine our goals. Our behavior is aimed at satisfying our needs, and so far as we can make out in this murky "no man's land" between biology, philosophy, and theology, the only consistent human purpose is for survival—of the individual and the species.

The conclusion that our motivation lacks direct sources in ratio-

---

* *The Importance of Being Earnest* (1895), Act I.

nality is not a counsel of despair, any more than the realization that science, the practical expression of reason, lacks the ability to direct its own application. By recognizing the disconnection between motivation and reason, we sharpen our inquiries and deliberations on policy in human affairs. If we look for the separation between reason and motivation, we will find it easier to understand our failures. We should ask, "Why do we want what we want?" as well as "How do we get what we want?" Reason has made an indispensable contribution to human success, and it alone is capable of defining its own limitations. As we learn these limitations we can build a framework within which we can place the knowable laws of nature and recognize our responsibilities as the most powerful species on the planet.

THE LIMITS OF THE MIND

Gunther Stent suggests that we have innate mental concepts that match the outside world, because they helped our ancient ancestors to survive over the course of evolution (1). This means that we should not expect the human mind to be able to comprehend everything in the universe—only what has been important for our past survival. For example, we find the concept of number helpful and easy to manage, so long as we are operating in the range of normal daily usage. We can deal with tens and even hundreds without too much difficulty, but we have trouble in grasping the meaning of huge numbers. The problem is illustrated by a sketch of numerical and biological perspective by Robert Pollack:

> We are very big compared to a cell, and cells are very big compared to the atoms of which they, and we, are composed. We are made of about 100 million million, or 10^{14} cells, and each cell is made up of 10^{14} atoms. Put another way, the complexity of the cell in molecular terms is about as great as the complexity of a person—brain and all—in cellular terms. This is as hopelessly unintuitive as the fact that the universe is hundreds of millions of times older—and

life on earth is tens of millions of times older—than the oldest living person (2).

These numbers are highly meaningful, but it is hard for our minds to grasp what they are telling us. A similar challenge can be found in quantum mechanics, where concepts can defy placement in our mental picture of the world. For example, we can define the laws that govern nuclear physics and we can harness atomic power, but we cannot really comprehend the underlying principles—in some settings we describe light in terms of waves, while in other settings we describe light in terms of particles. How can light be both waves and particles? The paradox cannot be reconciled; the mind is shaped by its evolutionary origins, and so it should come as no surprise that it has substantial limitations.

PROGRESS IS ADAPTATION

Tracing the story of reason through the 200,000 years of human pre-literate and literate history, an intelligible though often surprising story has unfolded. But from a broad perspective there is a paradox. People who study the history of human society claim to see "progress," while those who study the rest of the animal kingdom see only improving adjustments to the chances of survival in an unstable environment. This dissonance between human and biological history is more apparent than real, for what we call human progress is the product of resolute, unrealistic optimism—that our latest way of life is superior to anything that came before. Yet there is the indisputable fact that the twentieth century outperformed all others on the scale of horror—genocide in Armenia and Germany, barbaric civil wars in Russia, Spain, and China, with more recent massacres in Chile, Argentina, Cambodia, Rwanda, and Bosnia. In the First World War some 8.5 million soldiers died. The massive bloodshed from 1939 to 1945, estimated at 35 to 60 million deaths, primarily affected civilians, and the figure is imprecise because we could not keep track of how many died in Russia and China. In the face of such appalling events, we must conclude that *Homo sapiens* is not progressing any-where—if progress is taken to mean movement toward a better form of human being or human society.

The evolution of our species was a long and hazardous journey. Our ancestors had to struggle against forces beyond their control, but now we ourselves have become the major threat to our own survival. Since we have the ability to destroy each other so easily, the argument presented here—that progress is continuous adaptation—emerges as a statement that is neither destructive nor nihilistic. One theme of this book is that adjustment to promote survival of our species is a worthy and by no means trivial task.

It is hard to escape from our cultural prejudices when we try to interpret our own place in history, but if we look at human progress with a dispassionate and rational eye, all we simply find is *Homo sapiens,* like other animals, searching for the best way to adapt to changing circumstances. So far as we know, the structure of human brains has been the same since our species first appeared, and there is nothing to suggest that the function of human minds has changed—beyond having access to an expanded cultural "database" of experience. Our observations and interpretations have been checked, refined and extended over thousands of years, and gradu-ally much of the folklore has been replaced by knowledge concern-ing the laws of nature. It is this information that lets us do so much more than our ancestors. The accumulation of knowledge gives us power, but power for what? Our needs are the same as they have al-ways been, and our goals are to satisfy these needs. "Progress" implies something more—the notion of advancing along a pathway to reach some purposive end. Yet it is difficult to find rational support for this image.

Cultures can exert an influence on what kind of survival is tar-geted—more people or longer life—but there are limits. We can ex-pect a finite ceiling on how many people our planet can support and how long we can extend human life. The major gains of modern medicine have been alleviation of disease in infants, children, and young adults. We all have biological clocks ticking—the aging process entails wear and tear, and a declining capacity for repair. As individuals, we have built-in biological obsolescence that sustains the vigor of the species. Each generation must make room for the next, so that our genes can be reshuffled and future generations will have the strength to continue.

RELIGION AND SKEPTICISM

The debate between religion and skepticism will go on forever. Religion has the power of emotional warmth, while skepticism has the force of cold consistency. When emotions come into conflict with reason, emotions win—this is why most people in the world follow a religion. Cultures and religions tend to support each other, yet human culture has also developed in ways that allow people to reject their traditional religious inculcation; instead they seek enlightenment independently. Such people have pondered the possibility of something between the two extremes of conventional religion and atheism. Humanists such as Cyril Bibby stake out a claim for this middle ground:

> Humanism will challenge all religions in the form of demonology or miracle-mongering or unhistorical assertion. But it will recognize that religion sometimes has the aspects of wonder in the face of the great mysteries of the universe, of spiritual aspirations, of humility: and such religious attitudes will also be its own. It will, in short, seek to live by the text, "And ye shall know the truth, and the truth shall make you free." (3)*

Bertrand Russell has also championed the cause for seeking a middle ground between dogmatism and skepticism.

---

* The truth may or may not make us happy, but as we learn more about the laws that govern the world we can live longer, do more, and dismiss our illusions.

The sceptic says "nothing can be known"; he is a dogmatist, though a negative one. His creed, we must admit, is paralyzing, and a nation which accepts it is doomed to defeat, since it cannot adduce adequate motives for self-defense. But the scientific attitude is quite different. It does not say "knowledge is impossible," but "knowledge is difficult". . . . . The dogmatist accepts one hypothesis regardless of the evidence; the sceptic rejects all hypotheses regardless of the evidence. Both are irrational. The rational man accepts the most probable hypothesis for the time being, while continuing to look for new evidence to confute it.*

In chapter 8 we cited Planck and Einstein as experts in the use of reason who chose to embrace a form of religion, but theirs was not a conventional religion. Planck was the grandson of a Protestant pastor, and while he had some deep religious feelings he did not believe in "a personal God, let alone a Christian God" *(4).* Einstein had a Jewish background but his position was much the same. In mid-life, filling out bureaucratic forms in Switzerland, he described himself as "without religious denomination" *(5).* In 1929 he was asked directly: "Do you believe in God?" He replied: "I believe in Spinoza's God who reveals himself in the orderly harmony of what exists, not in a God who concerns himself with the fates and actions of human beings" *(6).* Later he wrote: "I have not found a better expression than 'religious' for the trust in the rational nature of reality that is, at least to a certain extent, accessible to human reason" *(6).* So Einstein and Planck both occupied the middle ground between dogmatism and skepticism.

A LIFE OF REASON
Philosophers from Socrates to Santayana have proclaimed that we should each try to hold on to a life of reason. This book echoes the

---

* Russell's words are as cogent as they are eloquent. This passage was first published by *The Listener* in 1948 and then quoted by Hector Hawton in *The Feast of Unreason* (London: Watts, 1952).

sentiment that we should strive to be as rational as possible, but it also sounds a cautionary note—that irrational forces drive motivation and underlie the quick responses we call "gut reactions." The purely rational human being, whose thought and behavior are the crystallization of absolute reason, is a fictional character who can never exist in the real world.

Why cannot reason direct our goals? The answer comes from our examination of reason's nature. Unlike emotions, reason does not entail needs that crave satisfaction. It is, furthermore, hard to imagine how reason would operate if it did crave satisfaction, for then it would not *compete* with emotions, it would *be* an emotion—we would *feel* reason in the way we feel anger (which craves a fight) or fear (which craves a flight). The separation of reason from motivation is fundamental to—even constitutive of—human cognition.

FROM THE DIFFERENT natures of reason and motivation, it follows that we must link the two together, for the capacity to solve problems is of little value without motivation, and vice versa. If we want to preserve ourselves and the environment, we need rational methods for dealing with global pollution, deforestation, and overpopulation, but rational methods are not enough. Wanting to deal with long-term challenges must prevail over wanting to ignore them for short-term gains. The hierarchy of our motives—establishing our priorities and choosing what we want *most*—is crucial. That hierarchy, this book argues, is within reason's domain to pursue, but beyond reason's ability to decide.

WORKS CITED

1: *Introduction*

1. Francis Bacon. *Novum Organum* (1620). Chicago: Open Court, 1994.
2. Sophocles. *Antigone.* Translated by H. Lloyd Jones. Cambridge, Mass.: Harvard University Press, 1994.
3. Aristotle. *Nicomachean Ethics.* Indianapolis: Bobbs-Merrill, 1980.
4. Cicero. *Tusculan Disputations.* Cambridge, Mass.: Harvard University Press, 1989.
5. V. A. Smith. *Akbar 1560–1605.* Oxford: Oxford University Press, 1919.
6. William Shakespeare. *Hamlet* (1601).
7. N. Hampson. *The Enlightenment.* Harmondsworth, England: Penguin Books, 1968.
8. Baruch Spinoza. *Tractatus theologico-politicus* (1670). Indianapolis: Hackett, 1991.

2: *A Definition of Reason*

1. Lesley Brown, ed. *The New Shorter Oxford English Dictionary.* Oxford: Clarendon Press, 1993.
2. Nicholas Rescher. *Rationality: A Philosophical Inquiry into the Nature and the Rationale of Reason.* Oxford: Clarendon Press, 1988.
3. J. D. Duffy and J. J. Campbell. "The Regional Prefrontal Syndromes: A Theoretical and Clinical Overview." *Journal of Neuropsychiatry and Clinical Neurosciences* 6 (1994): 379–387.
4. T. W. Settle. "Galileo's Knife." In J. Agassi and I. C. Jarvie, eds., *Rationality: The Critical View,* 181–200. Dordecht: Martinus Nijhoff, 1987.

3: *Language and Speech*

1. Gilbert Ryle. *The Concept of Mind.* London: Penguin Books, 1949.
2. Samuel Johnson. *Lives of the Most Eminent English Poets* (1779). Oxford: Oxford University Press, 1968.
3. Colin McGinn. *The Character of Mind.* Oxford: Oxford University Press, 1982.
4. J. J. Jenkins. "Language and Thought." In J. F. Voss, ed., *Approaches to Thought,* 211–237. Columbus, Ohio: Merrill, 1969.

5. M. Müller. *The Science of Thought.* New York: Scribner, 1887.

6. Hannah Arendt. *The Life of the Mind.* San Diego: Harcourt and Brace, 1981.

7. Ludwig Wittgenstein. *Tractatus Logico-Philosophicus* (1922). Translated by D. F. Pears and B. F. McGuiness. New York: Routledge, 1994.

8. D. F. Benson. *The Neurology of Thinking.* New York: Oxford University Press, 1994.

9. Steven Pinker. *The Language Instinct.* New York: William Morrow, 1995.

10. Oliver Sacks. *The Man Who Mistook His Wife for a Hat.* New York: Summit Books, 1970.

11. Zeno Vendler. "Meaning in Linguistics," under "Linguistics," in The New Encyclopaedia Britannica, 15th ed. (1997), 23:40–71.

12. Hilary Putnam. *Reason, Truth and History.* Cambridge: Cambridge University Press, 1981.

13. Benjamin Lee Whorf. *Language, Thought, and Reality.* Cambridge, Mass.: MIT Press, 1956.

14. M. Gopnik. "Dysphasia in an Extended Family." *Nature* 344 (1990):715.

15. U. Bellugi, A. Bihrle, T. Jernigan, D. Trauner, and S. Doherty. "Neuropsychological, Neurological, and Neuroanatomical Profile of Williams' Syndrome." *American Journal of Medical Genetics* 6 (1991):115–125.

16. Carl B. Boyer. *A History of Mathematics.* New York: Wiley, 1968.

17. Bertrand Russell. *A History of Western Philosophy.* London: George Allen and Unwin, 1945.

18. David Premack and Ann J. Premack. "How 'Theory of Mind' Constrains Language and Communication." *Discussions in Neuroscience* 10 (1994):93–130.

19. Michael S. P. Gazzaniga. *Nature's Mind: The Biological Roots of Thinking, Emotions, Sexuality, Language, and Intelligence.* New York: Basic Books, 1992.

4: *Social Behavior*

1. Seneca. *Epistles,* 41,8. Cited in *Bartlett's Familiar Quotations.* Boston: Little, Brown, 1992.

2. Baruch Spinoza. *Ethics* (1677), Part III, proposition 35. Chicago: Encyclopaedia Britannica, 1952.

3. Nicholas Humphrey. *A History of the Mind.* London: Chatto and Windus, 1992.

4. Edward O. Wilson. *The Diversity of Life.* Cambridge, Mass.: Belknap Press of Harvard University, 1992.

5. Robert Trivers. *Social Evolution.* Menlo Park, Calif.: Benjamin Cummings, 1985.

6. Irenaus Eibl-Eibesfeldt. *Love and Hate: The Natural History of Behavior Patterns.* New York: Holt, Rinehart and Winston, 1972.

7. J. F. Wittenberger. *Animal Social Behavior.* Boston: Duxbury Press, 1981.

8. R. W. Byrne and A. Whiten, eds. *Machiavellian Intelligence: Social Expertise and the Evolution of Intelligence in Monkeys, Apes, and Humans.* Oxford: Clarendon Press, 1988.

9. Niccolò Machiavelli. *The Prince* (1532). New York: New American Library, 1952.

10. Emil Menzel. "A Group of Young Chimpanzees in a One-Acre Field: Leadership and Communication." In R. Byrne and A. Whiten, eds., *Machiavellian Intelligence,* 155–160. Oxford: Clarendon Press, 1988.

11. Frans de Waal. *Good Natured: The Origins of Right and Wrong in Humans and Other Animals.* Cambridge, Mass.: Harvard University Press, 1996.

12. Clifford Geertz. *The Interpretation of Cultures.* New York: Basic Books, 1973.

13. Stephen Jay Gould. *Full House: The Spread of Excellence from Plato to Darwin.* New York: Harmony Books, 1996.

14. S. G. F. Brandon. *Man and His Destiny in the Great Religions.* Toronto: University of Toronto Press, 1962.

15. M. E. Thurston. *The Lost History of the Canine Race.* New York: Avon Books, 1996.

16. Alfred North Whitehead. *Science and the Modern World.* Cambridge: Cambridge University Press, 1932.

17. A. Rosman and P. G. Rubel. *The Tapestry of Culture.* New York: McGraw-Hill, 1981.

18. Ruth Benedict. *The Chrysanthemum and the Sword: Patterns of Japanese Culture.* Boston: Houghton Mifflin, 1946.

19. Donald E. Brown. *Human Universals.* New York: McGraw-Hill, 1991.

20. Denis Warner and Peggy Warner. *The Sacred Warriors: Japan's Suicide Legions.* New York: Van Nostrand Reinhold, 1982.

21. John Carey, ed. *The Faber Book of Reportage.* London: Faber and Faber, 1987.

22. Antonio R. Damasio. *Descartes' Error: Emotion, Reason, and the Human Brain.* New York: Putnam, 1994.

23. Stanley Garn. *Culture and the Direction of Human Evolution.* Detroit: Wayne State University Press, 1964.

24. A. Bechara, H. Damasio, D. Tranel, and S. W. Anderson. "Dissociation of Working Memory from Decision Making within the Human Prefrontal Cortex." *Journal of Neuroscience* 18 (1998):428–437.

25. J. D. Duffy and J. J. Campbell. "The Regional Prefrontal Syndromes: A Theo-

retical and Clinical Overview." *Journal of Neuropsychiatry and Clinical Neurosciences* 6 (1994): 379–387.

26. M. S. Mega and J. L. Cummings. "Frontal-Subcortical Circuits and Neuropsychiatric Disorders." *Journal of Neuropsychiatry and Clinical Neurosciences* 6 (1994): 358–370.

5: *Ethics*

1. Geoffrey J. Warnock. *The Object of Morality.* London: Methuen, 1971.
2. P. H. Nowell-Smith. *Ethics.* Harmondsworth, England: Penguin Books, 1954.
3. Frans de Waal. *Good Natured: The Origins of Right and Wrong in Humans and Other Animals.* Cambridge, Mass.: Harvard University Press, 1996.
4. Bertrand Russell. *A History of Western Philosophy.* London: George Allen and Unwin, 1945.
5. Epicurus. "Letter to Menoeceus." In *Hellenistic Philosophy,* ed. A. A. Long. London: Duckworth, 1995.
6. Michael Philips. *Between Universalism and Skepticism.* New York: Oxford University Press, 1994.
7. Sissela Bok. *Lying: Moral Choice in Public and Private Life.* New York: Vintage, 1989.
8. H. Cleckley. *The Mask of Sanity.* New York: New American Library, 1982.
9. Robert M. Hare. *Without Conscience: The Disturbing World of the Psychopaths among Us.* New York: Simon and Schuster, 1993.

6: *Commerce*

1. Frans de Waal. *Good Natured: The Origins of Right and Wrong in Humans and Other Animals.* Cambridge, Mass.: Harvard University Press, 1996.
2. John Stuart Mill. *Principles of Political Economy* (1871). W. J. Ashley, ed. London: Longmans, Green, 1909.
3. A. Hingston Quiggin. *A Survey of Primitive Money.* London: Methuen, 1949.
4. P. Einzig. *Primitive Money in Its Ethnological, Historical and Economical Aspects.* Oxford: Pergamon, 1966.
5. George Bernard Shaw. *The Irrational Knot.* New York: Brentano, 1905.
6. George Bernard Shaw. *Major Barbara.* New York: Brentano, 1907.
7. Paul Kennedy. *The Rise and Fall of the Great Powers.* New York: Random House, 1987.
8. A. S. Atiya. *Crusade, Commerce and Culture.* Bloomington: Indiana University Press, 1962.
9. H. Pirenne. *Early Democracies in the Low Countries.* New York: Norton, 1971.

10. N. Rosenberg and L. E. Birdzell. *How the West Grew Rich.* New York: Basic Books, 1986.

11. Robert L. Heilbroner. *The Worldly Philosophers.* New York: Simon and Schuster, 1980.

12. Adam Smith. *The Wealth of Nations* (1776). Buffalo: Prometheus Books, 1991.

13. John Maynard Keynes. *The General Theory of Employment, Interest and Money.* London: Macmillan, 1936.

14. J. Jacobs. *Systems of Survival.* New York: Random House, 1992.

15. P. Drucker. *Post-Capitalist Society.* New York: HarperCollins, 1993.

16. Noam Chomsky. "Equality: Language Development, Human Intelligence, and Social Organization." In *The Chomsky Reader,* ed. N. Chomsky and J. Peck, 183–202. New York: Pantheon Books, 1987.

17. B. J. McCormick. P. D. Kitchin, G. P. Marshall, A. A. Sampson, and R. Sedgwick. *Introducing Economics.* Harmondsworth, England: Penguin Books, 1983.

18. John Kenneth Galbraith. *The Good Society: The Humane Agenda.* Boston: Houghton Mifflin, 1996.

19. Adam Smith. *The Theory of Moral Sentiments* (1759). New York: Garland, 1971.

20. Frank Hahn. "Benevolence." In *Thoughtful Economic Man,* ed. J. G. T. Meeks. Cambridge: Cambridge University Press, 1991.

21. Paul Kennedy. *Preparing for the Twenty-First Century.* London: HarperCollins, 1993.

22. Joseph Schumpeter. *Capitalism, Socialism, and Democracy.* New York: Harper, 1950.

23. Charles Kindleberger. *Maniacs, Panics and Crashes.* New York: Basic Books, 1978.

24. Robert H. Frank. *Passions within Reason.* New York: Norton, 1988.

25. Jon Elster. *Ulysses and the Sirens.* Cambridge: Cambridge University Press, 1979.

26. R. Axelrod. *The Evolution of Cooperation.* New York: Basic Books, 1984.

27. R. Dawkins. *Nice Guys Finish First.* London: BBC Enterprises (Horizon Videocassette), 1986.

28. R. Keyes. *Nice Guys Finish Seventh.* New York: HarperCollins, 1992.

7: Government

1. Niccolò Machiavelli. *The Prince* (1532). New York: New American Library, 1952.

2. Thomas Smith. *De republica Anglorum; A Discourse on the Commonwealth of England* (1583). Cambridge: Cambridge University Press, 1906.

3. N. Hampson. *The Enlightenment.* Harmondsworth, England: Penguin Books, 1968.

4. Baron de Montesquieu. *L'Esprit des lois* (1748). Chicago: Encyclopaedia Britannica, 1952.

5. John Locke. *Second Treatise of Civil Government* (1690). Buffalo: Prometheus Books, 1986.

6. Edmund Burke. *Reflections on the Revolution in France.* Indianapolis: Hackett, 1987.

7. John Rawls. *Political Liberalism.* New York: Columbia University Press, 1993.

8. D. Martin. *General Amin.* London: Faber and Faber, 1974.

9. J. C. Fest. *Hitler.* Translated by Richard and Clara Winston. New York: Harcourt Brace Jovanovich, 1974.

10. Sebastian Haffner. *The Meaning of Hitler.* Translated by Ewald Osers. London: Weidenfeld and Nicolson, 1979.

11. D. Cameron Watt. *How War Came: The Immediate Origins of the Second World War, 1938–1939.* New York: Pantheon Books, 1989.

12. Robert Gellately. *The Gestapo and German Society: Enforcing Racial Policy, 1933–1945.* Oxford: Clarendon Press, 1990.

13. D. J. Goldhagen. *Hitler's Willing Executioners.* New York: Alfred A. Knopf, 1996.

14. J. Carey. *The Intellectuals and the Masses.* London: Faber and Faber, 1992.

15. B. Ljunggren. *Great Men with Sick Brains and Other Essays.* Park Ridge, Ill.: American Association of Neurological Surgeons, 1990.

8: *Reason for Religion*

1. Melford E. Spiro. "Religion: Problems of Definition and Explanation." In M. Banton, ed., *Anthropological Approaches to the Study of Religion,* 85–125. New York: Praeger, 1966.

2. Paul Johnson. *A History of the Jews.* London: Weidenfeld and Nicolson, 1987.

3. Bertrand Russell. *A History of Western Philosophy.* London: George Allen and Unwin, 1945.

4. Abraham H. Maslow. *Religions, Values, and Peak-Experiences.* New York: Viking Press, 1970.

5. S. G. F. Brandon. *Man and His Destiny in the Great Religions.* Toronto: University of Toronto Press, 1962.

6. E. B. Tylor. *Primitive Culture* (1871). Reprinted as *Religion in Primitive Culture.* New York: Harper and Row, 1958.

7. Edwin O. James. *Prehistoric Religion.* London: Thames and Hudson, 1957.

8. Huston Smith. *The World's Religions.* San Francisco: HarperCollins, 1991.

9. Claude Lévi-Strauss. *The Raw and the Cooked: Introduction to a Science of Mythology.* Translated by John and Doreen Weightman. New York: Harper and Row, 1969.

10. S. N. Eisenstadt, ed. *The Origins and Diversity of Axial Age Civilizations.* New York: State University of New York Press, 1986.

11. Plato. *Phaedo.* Translated by G. M. A. Grube. Indianapolis: Hackett, 1992.

12. Rudolf Otto. *The Idea of the Holy.* Translated by John W. Harvey. London: Oxford University Press, 1950.

13. Tlakaelel. Talk at Interface, Watertown, Mass., April 15, 1988. Quoted in M. P. Fisher, *Living Religions.* Englewood Cliffs, N.J.: Prentice Hall, 1994.

14. William James. *The Varieties of Religious Experience* (1902). Boston: Harvard University Press, 1985.

15. Arthur Koestler. *The Lotus and the Robot.* London: Hutchinson, 1960.

16. W. N. Pahnke. "Drugs and Mysticism." *International Journal of Parapsychology* 8 (1966):295–314.

17. Rita Dove. "On the Road to Damascus." In R. Eastman, ed., *The Ways of Religion: An Introduction to the Major Traditions.* 2d ed. New York: Oxford University Press, 1993.

18. K. L. Reichelt. *Meditation and Piety in the Far East: A Religious-Psychological Study.* Translated by S. Holth. New York: Harper, 1954.

19. G. Schüttler. *Die Letzen Tibeischen Orakelpriester: Psychiatrische-neurologische Aspekte.* Wiesbaden: Franz Steiner, 1971.

20. Edward O. Wilson. *On Human Nature.* Cambridge, Mass.: Harvard University Press, 1978.

21. Francis Bacon. "Of Atheism" (1625). In *Essays.* London: Oxford University Press, 1966.

22. Isaac Newton. *Opticks* (1730). New York: Dover, 1952.

9: *Reason Against Religion*

1. David Hume. "Of Superstition and Enthusiasm" (1741). In E. F. Miller, ed., *Essays Moral, Political and Literary.* Indianapolis: Library Classics, 1987.

2. Ambrose Bierce. *The Devil's Dictionary* (1881). New York: Dover, 1993.

3. H. L. Mencken. "Coda." In B. Rascoe, ed., *Smart Set.* New York: Reynal and Hitchcock, 1934.

4. Aldous Huxley. *Texts and Pretexts.* London: Chatto and Windus, 1949.

5. D. Carrasco. "Religions of Mesoamerica." In B. N. Earhart, ed., *Religious Traditions of the World.* San Francisco: HarperCollins, 1993.

6. Terry Jones and Alan Ereira. *Crusades.* London: BBC Enterprises, 1994.

7. A. S. Atiya. *Crusade, Commerce and Culture.* Bloomington: Indiana University Press, 1962.

8. Stephen Neill. *Colonialism and Christian Missions.* London: Butterworth, 1966.

9. Barbara W. Tuchman. *Bible and Sword.* New York: New York University Press, 1956.

10. Lecompte De Noüy. *The Road to Reason.* New York: Longmans Green, 1949.

11. M. Sharratt. *Galileo: Decisive Innovator.* Cambridge: Cambridge University Press, 1994.

12. H. Kamen. *The Spanish Inquisition.* New York: New American Library, 1965.

13. Barbara Rosen. *Witchcraft.* London: Edward Arnold, 1969.

14. H. R. Trevor-Roper. *The European Witch-Craze of the 16th and 17th Centuries.* London: Penguin Books, 1969.

15. M. Harris. *Culture, People, Nature: An Introduction to General Anthropology.* 4th ed. New York: Harper and Row, 1985.

16. A. M. Ludwig. "Altered States of Consciousness." *Archives of General Psychiatry* 15 (1966):225–234.

17. P. J. McKenna, J. M. Kane, and K. Parrish. "Psychotic Syndromes in Epilepsy." *American Journal of Psychiatry* 142 (1985):895–904.

18. E. Wyllie. *The Treatment of Epilepsy: Principles and Practice.* 2d ed. Baltimore: Williams and Wilkins, 1997.

19. William James. *The Varieties of Religious Experience* (1902). Boston: Harvard University Press, 1985.

20. David M. Wulff. *Psychology of Religion: Classic and Contemporary Views.* New York: John Wiley, 1991.

21. Michael A. Persinger. "Vectorial Cerebral Hemisphericity as Differential Sources for the Sensed Presence, Mystical Experiences and Religious Conversions." *Perceptual and Motor Skills* 76 (1993):915–930.

22. G. Sedman. "Being an Epileptic: A Phenomenological Study of Epileptic Experiences." *Psychiatria et Neurologia, Basel* 152 (1966):1–16.

23. K. Dewhurst and A. W. Beard. "Sudden Religious Conversions in Temporal Lobe Epilepsy." *British Journal of Psychiatry* 117 (1970):497–507.

24. S. Lockhart. "Joan of Arc: a psychiatric viewpoint." *Oxford Medical School Gazette* 31 (1980):37–40.

25. L. Perez. "The Messianic Psychotic Patient." *Israel Annals of Psychiatry and Related Disciplines* 14 (1977):364–373.

10: *Art*

1. Rebecca West. "The Strange Necessity" (1928). In *Rebecca West: A Celebration*. New York: Viking, 1977.

2. Ellen Dissanayake. *Homo Aestheticus: Where Art Comes From and Why*. Seattle: University of Washington Press, 1992.

3. Roger Scruton. *Modern Philosophy: An Introduction and Survey*. London: Mandarin, 1996.

4. R. Goldwater and Marco Treves. *Artists on Art from the XIV to the XX Century*. New York: Pantheon Books, 1972.

5. A. Wood. Revised by J. M. Bowsher. *The Physics of Music*. London: Methuen, 1969.

6. John Carey, ed. *The Faber Book of Reportage*. London: Faber and Faber, 1987.

7. I. C. Jarvie. "The Objectivity of Criticism of the Arts." In J. Agassi and I. C. Jarvie, eds., *Rationality: The Critical View*, 201–216. Dordrecht: Martinius Nijhoff, 1987.

8. E. H. Gombrich. *The Story of Art*. 4th ed. London: Phaidon, 1951.

9. M. Baxandall. *Painting and Experience in Fifteenth-Century Italy*. 2d ed. Oxford: Oxford University Press, 1988.

10. R. A. Henson and M. Critchley, eds. *Music and the Brain*. London: Heinemann, 1977.

11. Terry Eagleton. *Literary Theory: An Introduction*. Oxford: Blackwell, 1983.

12. E. M. Zuesse. "Ritual." In M. Eliade, ed., *The Encyclopedia of Religion*, 405–422. New York: Macmillan, 1987.

11: *Science*

1. Herbert Butterfield. "Dante's View of the Universe." In *The History of Science*, 15–24. London: Cohen and West, 1951.

2. M. Mulkay. *Science and the Sociology of Knowledge*. London: Allen and Unwin, 1979.

3. Bertrand Russell. *The Impact of Science on Society*. New York: Columbia University Press, 1951.

4. Francis Bacon. "Meditationes Sacrae, De Haeresibus. Proposition 35" (1597). In *Works of Francis Bacon*. New York: Garrett Press, 1968.

5. Thomas H. Huxley. "The Method of Zadig." In *Collected Essays* (1893–1894). London: Macmillan, 1895.

6. Thomas H. Huxley. *Lay Sermons, Addresses, and Reviews—A Liberal Education*. New York: Appleton, 1882.

7. Alfred North Whitehead. *Science and the Modern World.* Cambridge: Cambridge University Press, 1932.

8. E. Borel. *Le hasard.* Paris: Presses Universitaires de France, 1948.

9. Francis Bacon. *The Advancement of Learning* (1605). London: J. M. Dent, 1934.

10. Karl R. Popper. *The Logic of Scientific Discovery.* London: Hutchinson, 1959.

11. Albert Einstein and L. Infeld. *The Evolution of Physics: The Growth of Ideas from Early Concepts to Relativity and Quanta.* New York: Simon and Schuster, 1938.

12. H. Blaschko. "Metabolism and Storage of Biogenic Amines." *Experientia (Basel)* 13 (1957):9–12.

13. H. Ehringer, H. Oleh Hornykiewicz. "Distribution of Noradrenaline and Dopamine (3-Hydroxytyramine) in the Human Brain and their Behavior in Diseases of the Extrapyramidal System." *Klinische Wochenschrift* 38:1236–1239.

14. Thomas H. Huxley. "Biogenesis and Abiogenesis." In *Collected Essays* (1893–1894). London: Macmillan, 1895.

15. P. B. Medawar. *Advice to a Young Scientist.* New York: Harper and Row, 1979.

16. L. Ordenstein. "Sur la paralysie agitante et la sclérose en plaques généralisée." M.D. thesis. Paris: Martinet, 1867.

17. H. Corrodi, K. Fuxe, T. Hokefelt, P. Lidbrink, and U. Ungerstedt. "Effect of Ergot Drugs on Central Catecholamine Neurons: Evidence for a Stimulation of Central Dopamine Neurons." *Journal of Pharmacy and Pharmacology* 25 (1973):178–183.

18. Hilary Putnam. "Philosophers and Human Understanding." In A. F. Heath, ed., *Scientific Explanation,* 99–120. Oxford: Clarendon Press, 1981.

19. Karl R. Popper. *The Open Society and Its Enemies,* vol. 1, 5th ed. London: Routledge and Kegan Paul, 1966.

20. Albert Einstein. *Ideas and Opinions* (based on *Mein Weltbild*). New York: Crown, 1954.

21. G. Holton. *Science and Anti-Science.* Cambridge, Mass.: Harvard University Press, 1993.

22. Alan D. Sokal. "Transgressing the Boundaries—Toward a Transformative Hermeneutics of Quantum Gravity." *Social Text* (1996):217–252.

23. S. Weinberg. "Sokal's Hoax." *New York Review of Books* 43 (August 8, 1996):11–15.

24. D. F. Horrobin. "The Philosophical Basis of Peer Review and the Suppression of Innovation." *Journal of the American Medical Association* 263 (1990): 1438–1441.

25. B. Glick, I. S. Chang, and R. G. Jasp. "The Bursa of Fabricius and Antibody Production." *Poultry Science* 85 (1956):224–236.

26. H. A. Krebs. "The History of the Tricarboxylic Acid Cycle." *Perspectives. Biology and Medicine* 14 (1970):154–170.

27. R. S. Yalow. "Peer Review and Scientific Revolution." *Biological Psychiatry* 21 (1966):1–2.

12: *Behavior and the Brain*

1. D. Premack and G. Woodruff. "Does the Chimpanzee Have a Theory of Mind?" *Behavioural and Brain Sciences* 4 (1978):515–526.

2. K. Dunlap. "Are There Any Instincts?" *Journal of Abnormal Psychology* 14 (1919):35–50.

3. F. A. Beach. "The Descent of Instinct." *Psychological Review* 62 (1955):401–410.

4. A. N. Epstein. "Instinct and Motivation as Explanations for Complex Behaviour." In D. W. Pfaff, ed., *The Physiological Mechanisms of Motivation,* 25–58. New York: Springer Verlag, 1981.

5. Robert H. Thouless. *General and Social Psychology.* 4th ed. London: University Tutorial Press, 1958.

6. Richard Dawkins. *The Selfish Gene.* Oxford: Oxford University Press, 1976.

7. Abraham H. Maslow. *Motivation and Personality.* New York: Harper, 1945.

8. J. Maynard Smith. *Evolution and the Theory of Games.* Cambridge: Cambridge University Press, 1982.

9. Paul Ekman. "Are There Basic Emotions?" *Psychological Review* 99 (1992): 550–553.

10. C. McGinn. *The Character of Mind.* Oxford: Oxford University Press, 1982.

11. William James. *The Principles of Psychology* (1890). Cambridge, Mass.: Harvard University Press, 1981.

12. R. de Sousa. *The Rationality of Emotion.* Cambridge, Mass.: MIT Press, 1987.

13. A. R. Damasio. *Descartes' Error: Emotion, Reason, and the Human Brain.* New York: Putnam, 1994.

14. D. Goleman. *Emotional Intelligence.* New York: Bantam Books, 1995.

15. R. G. M. Heath. "Pleasure and Brain Activity in Man. Deep and Surface Electroencephalograms During Orgasm." *Journal of Nervous and Mental Diseases* 154 (1972):3–18.

16. M. A. Persinger. "Vectorial Cerebral Hemisphericity as Differential Sources for the Sensed Presence, Mystical Experiences and Religious Conversions." *Perceptual and Motor Skills* 76 (1993):915–930.

17. D. M. Tucker. "Lateral Brain Function, Emotion, and Conceptualization." *Psychological Bulletin* 89 (1981):19–46.

18. R. C. Bolles. *Theory of Motivation,* 2d ed. New York: Harper and Row, 1975.

19. A. Bain. *The Senses and the Intellect.* London: Longmans Green, 1868.

20. Gerald M. Edelman. *Neural Darwinism: The Theory of Neuronal Group Selection.* New York: Basic Books, 1987.

21. Thomas Carlyle. *Sartor Resartus* (1834). Oxford: Oxford University Press, 1987.

22. D. M. Waddle. "Matrix Correlation Tests Support a Single Origin for Modern Humans." *Nature* 368 (1994):452–454.

23. Charles Darwin. *The Descent of Man and Selection in Relation to Sex* (1871).

24. Clifford Geertz. *The Interpretation of Cultures.* New York: Basic Books, 1973.

25. M. Barinaga. "New Clue to Brain Wiring Mystery." *Science* 270 (1995):581.

26. Patrick Doherty, M. S. Fazeli, and F. S. Walsh. "The Neural Cell Adhesion Molecule and Synaptic Plasticity." *Journal of Neurobiology* 26:437–446.

27. C. S. Sherrington. *Man on His Nature.* Cambridge: Cambridge University Press, 1940.

28. Michael S. Gazzaniga. *The Mind's Past.* Berkeley: University of California Press, 1998.

29. G. R. Fink, P. W. Halligan, J. C. Marshall, C. D. Frith, R.S.J. Frakowiak, and R. J. Dolan. "Where in the brain does visual attention select the forest and the trees?" *Nature* 382 (1996): 626–628.

13: *Mind*

1. Plato. *Phaedo.* Translated by G. M. A. Grube. Indianapolis: Hackett, 1992.

2. A. J. Ayer. *Language, Truth, and Logic.* London: Gollancz, 1936.

3. A. J. Warnock. "Saturday Mornings." In *Essays on J. L. Austin.* Oxford: Clarendon Press, 1973.

4. Baruch Spinoza. "Proposition (Mind and Brain)." In *Ethics,* Part II (1677). London: J. M. Dent, 1989.

5. H. Nakanishi, Y. Sun, R. Nakamura et al. "Positive Correlations Between Cerebral Protein Synthesis Rates and Deep Sleep in Macaca Mulatta." *European Journal of Neuroscience* 9 (1997):271–279.

6. P. J. Boyle, J. C. Scott, A. J. Krentz et al. "Diminished Brain Glucose Metabolism Is a Significant Determinant for Falling Rates of Systemic Glucose Utilization during Sleep in Normal Humans." *Journal of Clinical Investigation* 93 (1994):529–535.

7. William James. *The Principles of Psychology* (1890). Cambridge, Mass.: Harvard University Press, 1981.

8. Edward O. Wilson. *Consilience: The Unity of Knowledge.* New York: Alfred A. Knopf, 1998.

9. P. M. Churchland. *Matter and Consciousness.* Cambridge, Mass.: MIT Press, 1988.

10. Francis Crick. *The Astonishing Hypothesis: The Scientific Search for the Soul.* New York: Touchstone, 1995.

11. B. J. Baars. "Metaphors of Consciousness and Attention in the Brain." *Trends in Neurosciences* 21 (1998):58–64.

12. George O. Romanes. *Mental Evolution in Animals.* London: Kegan Paul, Trench, 1883.

13. John Locke. *An Essay Concerning Human Understanding* (1690). New York: Penguin Books, 1974.

14. Frans de Waal. *Good Natured: The Origins of Right and Wrong in Humans and Other Animals.* Cambridge, Mass.: Harvard University Press, 1996.

15. Santiago Ramón y Cajal. *Chácharas de Café* (1923). Translated by F. H. Garrison and published as a separate addendum to *Degeneration and Regeneration of the Nervous System.* Birmingham, Ala.: Classics of Neurology and Neurosurgery Library, 1984.

16. Colin McGinn. *The Character of Mind.* Oxford: Oxford University Press, 1982.

17. Nicholas A. Humphrey. *History of the Mind.* London: Chatto and Windus, 1992.

18. L. R. Squire. "Mechanisms of Memory." *Science* 232 (1995):1612–1619.

19. C. J. Duffy. "Implicit Memory, Knowledge without Awareness." *Neurology* 49 (1997):1200–1202.

20. M. Seeck, N. Mainwaring, R. Cosgrove et al. "Neurophysiological Correlates of Implicit Face Memory in Intracranial Visual Evoked Potentials." *Neurology* 49 (1997):1312–1316.

21. S. B. Floresco, J. K. Seamans, and A. G. Phillips. "Selective Roles for Hippocampal, Prefrontal Cortical, and Ventral Striatal Circuits in Radial-Arm Maze Tasks With or Without a Delay." *Journal of Neuroscience* 17 (1997): 1880–1890.

22. J. R. Stellar and E. Stellar. *The Neurobiology of Motivation and Reward.* New York: Springer Verlag, 1985.

23. David A. Hume. *A Treatise of Human Nature* (1739). Harmondsworth, England: Penguin Books, 1987.

24. D. F. Norton. *The Cambridge Companion to Hume.* Cambridge: Cambridge University Press, 1993.

25. Yoshihisa Kashima. "Culture, Narrative, and Human Motivation." In D. Munro, J. F. Schumaker, and S. Carr, eds., *Motivation and Culture*, 16–30. New York: Routledge, 1997.

26. A. Bechara, H. Damasio, D. Tranel, and A. R. Damasio. "Deciding Advantageously Before Knowing the Advantageous Strategy." *Science* 275 (1997): 1293–1295.

27. D. F. Halpern. "Reasoning." In M. W. Eysenck, ed., *The Blackwell Dictionary of Cognitive Psychology*, 304–308. Oxford: Blackwell, 1990.

28. A. Tversky and D. Kahneman. "The Framing of Decisions and the Psychology of Choice." *Science* 211 (1981):453–458.

29. Stuart Sutherland. *Irrationality: The Enemy Within.* London: Penguin Books, 1994.

30. Stanley Milgram. *Obedience to Authority: An Experimental View.* New York: Harper and Row, 1969.

31. Thomas C. Schelling. *The Strategy of Conflict.* Cambridge, Mass.: Harvard University Press, 1960.

32. L. Weiskrantz. *Blindsight.* New York: Oxford University Press, 1986.

14: *Epilogue*

1. Gunther S. Stent. "Limits to the Scientific Understanding of Man." *Science* 187 (1975):1052–1057.

2. Robert Pollack. *Signs of Life: The Language and Meanings of DNA.* Boston: Houghton Mifflin, 1994.

3. Cyril Bibby. "Towards a Scientific Humanist Culture." In A. J. Ayer, ed., *The Humanist Outlook,* 13–27. London: Pemberton, 1968.

4. Max Planck to W. H. Kick, 18 June 1947, in F. Herneck, ed., *Wissenschaftsgeschichte: Vortrage und Abhandlugen,* 89–90. Berlin: Akademie-Verlag, 1984.

5. A. Fösling. *Albert Einstein.* New York: Viking, 1997.

6. R. W. Clark. *Einstein: The Life and Times.* New York: Avon, 1972.

INDEX

ABOUT THE AUTHOR

DONALD B. CALNE is director of the Neurode-
generative Disorders Centre at Vancouver
Hospital and professor of neurology at the
University of British Columbia. In 1999 he
was appointed Officer of the Order of Canada
for pioneering research on diseases of the
brain. He lives in Vancouver.